Crafting the Future

The DIY Factory & New Work Nexus

by
Jordan Grey

Crafting the Future

The DIY Factory & New Work Nexus

Contents

Introduction:
Crafting a New Era of Work

In a world that's evolving faster than ever, the way we work and create is undergoing its most significant transformation in centuries. This revolution is not just about changing tools and technologies; it's about a fundamental shift in our approach to work, creativity, and production. The dawn of this new era brings with it an exciting merger of the Do-It-Yourself (DIY) ethos with forward-thinking work practices, commonly referred to as New Work. Together, they are crafting a future where work is not just a means to an end but a source of fulfillment, innovation, and connection.

The rise of the DIY culture and the principles of New Work aren't mere trends. They represent a move away from traditional models that dominated the industrial and early information ages, proposing instead a framework where autonomy, creativity, and technology converge to redefine what it means to produce and innovate. This shift isn't just happening in secluded pockets of high-tech or artistic communities; it's spreading across industries, reshaping offices, factories, and workshops worldwide.

At the heart of this movement is a belief that everyone has the potential to be a maker, an innovator, and an impactful contributor to a more sustainable and creative economy. By embracing the DIY mindset, individuals and organizations are breaking down the barriers between the creator and the consumer, turning passive audiences into active participants in the creation process. This change is empowering

1

people to solve real-world problems, reimagining how goods are produced, and services are rendered.

Meanwhile, New Work principles champion a similar ethos but on the organizational and systemic level. They advocate for work environments that prioritize flexibility, autonomy, and a sense of purpose. In these new paradigms, hierarchical structures give way to flat, agile teams. Success is measured not just by output but by the growth and satisfaction of both individuals and communities. This approach not only fosters innovation but also enhances well-being and work-life harmony.

The merger of the DIY ethos and New Work principles is not without its challenges. Transitioning from traditional business models and work habits to more fluid, decentralized, and creative ones requires a significant shift in mindset and operational design. However, the examples of those who have successfully navigated this journey are both inspiring and instructive, offering blueprints for others to follow.

This transformation also sets the stage for unprecedented innovation. By designing collaborative workspaces and leveraging new tools and technologies, we can create environments that encourage experimentation and creativity. These spaces become not just places of work but hubs of innovation where the future is constantly being reimagined and recreated.

The DIY factory model embodies this innovative spirit. It is a place where the barriers between different disciplines blur, fostering a culture of continuous learning, adaptation, and cross-pollination of ideas. The principles it upholds are not just for individual workshops but for industries at large, suggesting a future where businesses operate more like interconnected ecosystems than isolated entities.

Yet, the evolution toward this new era of work is more than just an economic imperative. It's a response to a deeper, more universal human need for creativity, purpose, and connection. In an age where automation and artificial intelligence are reshaping the landscape of employment, the DIY and New Work movements offer a vision of the future where technology augments human creativity rather than replaces it.

The implementation of these principles into everyday work life also has profound implications for leadership and organizational culture. It calls for leaders who can inspire innovation and empower their teams, fostering an environment where feedback is not just encouraged but celebrated. This cultural shift promotes not just productivity but also a deeper sense of belonging and community among employees.

Moreover, this new era is marked by a commitment to sustainability and social responsibility. By adopting more localized and decentralized production methods, DIY factories and New Work practices challenge the traditional take-make-waste model, paving the way for more sustainable and ethical business practices. This alignment of economic activities with environmental and social values represents a holistic view of success in the 21st century.

The movements also highlight the role of education and skill development in preparing for the future. They call for a rethinking of traditional educational models to cultivate creativity, innovation, and a DIY mindset from an early age. This approach not only prepares individuals for the jobs of the future but also empowers them to become lifelong learners and adaptors in an ever-changing world.

At the intersection of these movements lies a powerful opportunity to reshape our work culture and economy. But realizing this vision requires more than just adopting new tools or practices; it requires a collective rethink of our values, our goals, and how we define success. It challenges us to imagine a world where work is not just a means of survival but a source of joy, growth, and fulfillment.

The road to this new era will not be without its obstacles. Skepticism, resistance to change, and the inertia of established systems will be continual hurdles. Yet, the potential rewards – a more creative, sustainable, and equitable world – are too great to ignore. By embracing a mindset of experimentation, resilience, and collaboration, we can overcome these barriers.

The chapters that follow detail the principles, practices, and possibilities of this new era. They are both a guide and an invitation: a guide for navigating the changing landscape of work and an invitation to be part of a community of innovators shaping this new horizon. As we stand on the cusp of this transformative moment, let us seize the opportunity to craft a future of work that is not only productive but profoundly human.

Ready to embark on this journey? Let's begin by exploring the rise of the DIY ethos, its historical roots, and how it's merging with New Work principles to create a fertile ground for innovation and change.

Chapter 1:
The Rise of the DIY Ethos

In the heart of an evolving professional landscape brimming with potential, the DIY Ethos emerges as a beacon of empowerment and innovation. This ethos, rooted deeply in the conviction that anyone can generate significant value through creativity and resourcefulness, is transforming the way we approach work and entrepreneurship. The historical journey from traditional handicraft to the high-tech emboldens this narrative, illustrating the unyielding human spirit to create, adapt, and overcome. In today's digital age, where information and resources are at our fingertips, the DIY culture has found fertile ground, flourishing in communities and workplaces alike. It's not merely about making or repairing things yourself, but a profound shift towards taking ownership, challenging the status quo, and believing in the transformative power of individual action. As we peel back the layers of this movement, we uncover a rich tapestry of stories, methodologies, and tools that bridge the gap between the possible and the actualized. This chapter sets the stage for a journey into understanding how the DIY ethos is not just a trend, but a foundational element in crafting a new era of work, where innovation, self-reliance, and a pioneering spirit drive us towards a future teeming with untapped potentials.

The Historical Roots of DIY

The genesis of the DIY (Do It Yourself) movement traces back to a time when crafting by hand was not merely a hobby but a fundamental aspect of daily life. This ethos, deeply embedded in the human spirit, thrived on the principles of self-reliance, creativity, and the intrinsic satisfaction derived from creating something tangible. As the industrial revolution ushered in mass production, it seemed for a time that the personal touch in creation might be lost. However, rather than extinguishing this innate desire to create, the advancements in technology and the proliferation of digital tools have only served to rekindle it, propelling the DIY movement into new realms. Today, the spirit of DIY has evolved beyond simple crafts and repairs, embodying a powerful ethos of innovation, empowerment, and a proactive approach to problem-solving. It's a testament to the enduring human penchant for creativity and a reminder that in every individual lies a potential innovator. This foundational pillar not only harks back to the essence of human ingenuity but also aligns perfectly with the modern drive towards more personalized, meaningful, and sustainable lifestyles and work cultures. As we continue to navigate the complexities of the digital age, the historical roots of DIY serve as a beacon, guiding forward-thinking professionals, entrepreneurs, and enthusiasts toward embracing a more hands-on, creative approach in their endeavors and professions.

From Handicraft to High-Tech The journey from traditional handicrafts to high-tech manufacturing is a tale of evolution, reflecting our society's shifts towards a future where innovation and creativity reign supreme. This transformation, deeply rooted in the DIY ethos, is redefining what it means to produce, innovate, and create in the 21st century.

In the past, handicrafts were the backbone of production, embodying personal skill, craftsmanship, and a deep connection to the materials used. Artisans took pride in their work, with every item telling a story of tradition, culture, and human touch. This tactile, human-centric approach to creation is something that, in the exhilarating rush towards high-tech, we risk losing—unless we strive to integrate the values of craftsmanship into modern manufacturing practices.

Enter the age of high-tech manufacturing, where 3D printers, CNC machines, and robotics have taken center stage. These technologies have opened up new realms of possibility, enabling us to create complex, precise products with speeds and efficiency that craftsmen of the past could only dream of. Yet, they also pose a challenge: how do we maintain the soul of craftsmanship in a world dominated by machines?

The answer lies in the fusion of the DIY ethos with modern technology. This convergence is not just about leveraging technology for mass production; it's about rekindling the artisan spirit—creativity, innovation, and personalization—in the context of high-tech manufacturing. It's about viewing these powerful tools as extensions of the human hand and mind, enabling us to achieve greater heights of creativity and craftsmanship.

As we stand on this bridge between past and future, it's essential to recognize that the DIY ethos serves as the heartbeat of this new era. It encourages us to tinker, experiment, and innovate. It reminds us that making mistakes is a part of learning and growing. It urges us to think beyond the conventional, pushing the boundaries of what's possible.

This ethos, when applied to high-tech manufacturing, transforms the way we approach production. It's no longer just about what can be made, but about what should be made. It prompts us to ask important questions about sustainability, ethics, and the impact of our creations on society and the environment. In doing so, it invests modern manufacturing with a sense of purpose and responsibility that goes beyond mere production.

Consider the potential for customisation that technology brings. In a world tailored to individual needs and preferences, the line between producer and consumer blurs. Consumers become creators, participating in the design and customization of products. This level of engagement is reminiscent of the craftsman's connection to their creations, reestablished in the digital age through technology.

Moreover, the democratization of technology has lowered the barriers to entry for innovation. High-tech tools that were once accessible only to large corporations are now within reach of individuals and small startups. This shift is propelling a new wave of entrepreneurs and innovators who combine the DIY spirit with high-tech tools to bring unique products and solutions to the market.

Yet, embracing this blend of handicraft and high-tech isn't without its challenges. There's a steep learning curve, not just in mastering the technology, but in integrating it with the creativity, intuition, and flexibility that characterize artisanal work. Additionally, as production methods evolve, so too must our skills and ways of thinking. It requires a commitment to lifelong learning, openness to change, and a willingness to experiment and fail.

The impact of this transformation extends beyond the sphere of production; it's reshaping our entire work culture. New work paradigms, centered around creativity, innovation, and flexibility, are emerging. The principles of the New Work movement align closely with the ethos of DIY factories, celebrating autonomy, passion, and a multidisciplinary approach to projects.

Forward-thinking companies are beginning to recognize the value of bridging handicraft and high-tech. They are fostering environments where creativity and innovation flourish, where employees are empowered to experiment, and where traditional roles are reconceived to encourage a culture of continuous learning and collaborative problem-solving.

The synergy between handicraft and high-tech also raises important questions about the future of work. As automation and AI continue to advance, what place is there for the human touch in manufacturing? The answer lies in leveraging technology not to replace human creativity and skill but to augment and enhance it. It's about creating a future where technology serves as a tool for empowering artisans and makers, not rendering them obsolete.

This evolution from handicraft to high-tech doesn't spell the end for traditional craftsmanship; rather, it signals a new chapter where the values and principles that have long defined craftsmanship are infused into modern production processes. It represents an opportunity to revisit and reinvigorate the core of what it means to be a maker in the modern world.

In embracing this shift, we must ensure that the drive towards efficiency and scalability doesn't overshadow the fundamental human desire to create, innovate, and imbue our creations with meaning. As we navigate this complex landscape, one thing remains clear: the future of manufacturing and work culture lies in striking a balance between the rich heritage of handicraft and the boundless possibilities of high-tech.

Ultimately, the journey from handicraft to high-tech is not about leaving behind the past but about carrying its essence forward into the future. It's a call to action for innovators, entrepreneurs, and makers of all stripes to forge a new path, one that marries the wisdom of tradition with the innovation of technology. This fusion is the key to unlocking a future that is not only productive and innovative but also deeply fulfilling and human-centric.

DIY Culture in the Digital Age

The proliferation of digital technologies has markedly transformed the Do-It-Yourself (DIY) ethos from a niche hobbyist pursuit into a central component of contemporary work culture and innovation. In essence, the digital age has democratized the means of production, design, and creativity, making these tools more accessible than ever before. This shift is not just about the individual capability to create and innovate but is also fundamentally altering how professionals, entrepreneurs, and enthusiasts alike approach problem-solving and product development.

At the heart of this transformation is the internet, a vast repository of knowledge and a platform for unparalleled connectivity. From forums and social media groups dedicated to DIY projects to online marketplaces for sourcing materials, the digital age has facilitated a support system and infrastructure that allows even the most niche of projects to find its audience and necessary resources.

Moreover, digital fabrication technologies such as 3D printing, CNC machining, and laser cutting have lowered the barriers to entry for manufacturing physical products. These tools, once the exclusive domain of large industries, are now accessible in local makerspaces and even home garages—enabling a rapid prototyping revolution. The impact on the innovation ecosystem is profound, as individuals and small teams can iterate designs swiftly without the need for significant capital investment.

Software development, too, has seen a parallel shift with open-source platforms and collaborative tools fostering a culture of sharing and co-creation. This has allowed for a rapid acceleration in software product development, with communities coming together to contribute code, fix bugs, and add features to projects that they are passionate about.

The rise of platforms facilitating crowdfunding and the gig economy has further supported the DIY ethos in the digital age. Entrepreneurs can now raise funds for their projects from a global audience, and individuals can offer their skills on project-based work, enabling a flexibility and autonomy that was hard to imagine in the traditional 9-to-5 work arrangement.

This digital empowerment is reshaping not just how things are made but also who gets to make them. Diverse voices, who were previously sidelined in mainstream narratives of innovation due to lack of access or resources, are now able to contribute their perspectives, leading to richer, more varied solutions to problems and a more inclusive design process.

The environmental implications of this shift are also notable. The DIY culture, facilitated by digital tools, encourages the repair, repurposing, and upcycling of materials, contributing to a more sustainable and less waste-producing model of consumption. The ethos of 'making do' and 'fixing' rather than disposing and buying new is gaining ground, supported by online tutorials and communities dedicated to these practices.

Furthermore, the integration of DIY practices into education is preparing future generations for a rapidly changing world. Schools and universities are incorporating maker spaces, coding boot camps, and design thinking workshops into their curricula, recognizing the importance of hands-on learning and the development of problem-solving skills. This educational shift is not just about creating a workforce capable of navigating the future but also about instilling a sense of agency and empowerment in individuals to shape their world.

Yet, challenges remain. The digital divide still excludes many from participating fully in this DIY revolution. Moreover, the skills and knowledge necessary to leverage these technologies effectively are not uniformly distributed, creating a gap between potential and actual empowerment. As such, there is a significant role for both educational institutions and policymakers to ensure that access to these empowering tools and the skills to use them are broadly accessible.

In a broader sense, the rise of the DIY culture in the digital age signals a larger societal shift towards valuing creativity, innovation, and personal fulfillment. The traditional consumerist model, reliant on passive consumption, is being challenged by a more active, engaged form of participation in the economy and society.

For forward-thinking professionals, entrepreneurs, and enthusiasts, the opportunities presented by this shift are vast. By embracing and implementing the practices of the DIY culture within their professional environments, they can spur innovation, foster a sense of community and collaboration, and contribute to a more sustainable and inclusive future.

The digital age has thus not only provided the tools but also set the stage for a transformative cultural shift. As we stand at the cusp of this change, the question isn't so much about whether the DIY ethos will continue to grow— but rather, how we can harness this momentum to build a world that reflects our collective aspirations for creativity, autonomy, and shared prosperity.

In conclusion, the emergence of DIY culture in the digital age is a testament to the power of technology to empower individuals and communities. By leveraging digital tools and platforms, we can break down barriers, foster innovation, and create more inclusive and sustainable practices. As we navigate this shifting landscape, it is clear that the ethos of doing it yourself is not a fleeting trend but a robust approach to life and work in the 21st century.

Chapter 2:
Understanding New Work

As we delve into the essence of New Work, it's crucial to grasp that this is more than a fleeting trend; it's a foundational shift in how we perceive, engage with, and ultimately innovate within the contemporary work landscape. The transition from rigid, industrial models to dynamic, information-fueled environments marks a radical reimagining of productivity and creativity. The New Work movement, at its core, champions flexibility, autonomy, and the significance of aligning one's work with personal values and societal needs, thereby offering a blueprint for fulfillment and societal contribution through one's professional endeavors. Engaging with this movement requires not just an appreciation of its principles but a willingness to adapt and foster a culture where the confluence of diverse ideas and cutting-edge technology paves the way for unprecedented innovation and problem-solving. Through this exploration, we're not just observing a shift in work paradigms; we're participating in the cultivation of environments where work is not just what we do, but a reflection of our most genuine selves and our collective aspirations for a better future.

The Evolution of Work Paradigms

In a world where change is the only constant, the evolution of work paradigms has been particularly transformative, propelling us from structured, industrial routines into a dynamic, information-fueled era. This seismic shift isn't just about the tools we use; it's about reimagining the very essence of "work" and how we engage with it. Today's professionals are moving away from the one-size-fits-all approach, leaning into roles that blend their passions, skills, and the flexibility offered by technological advancements. This transition has sparked a revolution in workplace environments, management styles, and career trajectories. The nature of work has become more fluid, decentralized, and intertwined with our personal values and lifestyles, setting the stage for innovations like DIY factories and the New Work movement. Embracing this shift requires a bold rethinking of traditional paradigms, an openness to experimenting with new models, and a commitment to continual learning. It's about creating work that not only sustains us economically but fulfills us creatively and socially, forging paths that previous generations could barely imagine. As we dive deeper into understanding these emergent paradigms, let's keep our minds open to the boundless possibilities they present, nurturing a future where work is not just something we do but a reflection of who we are and what we aspire to be.

From Industrial to Information Age In the grand tapestry of human progress, the shift from the Industrial to the Information Age represents more than a mere evolution—it signifies a profound transformation in how we work, think, and live. This epochal change has ushered in a new era of opportunity, challenging us to rethink the very nature of labor, creativity, and collaboration.

During the Industrial Age, work was predominantly defined by manual labor, mass production, and a strict hierarchy of roles and responsibilities. Factories were the emblem of economic prowess, with assembly lines churning out products at an unprecedented scale. Yet, this era was also characterized by its rigid, often dehumanizing work environments, where the individual worker's creativity and autonomy were frequently suppressed in favor of efficiency and uniformity.

Crafting the Future: The DIY Factory & New Work Nexus

As we transitioned into the Information Age, the nature of work underwent a seismic shift. Technologies, particularly computing and the internet, emerged as the backbones of this new era, transforming how information is created, processed, and distributed. This digital revolution democratized knowledge, upended traditional economic structures, and sparked an unprecedented wave of innovation.

In the midst of this transformation, the digital age brought forth a new set of values concerning work and productivity. Flexibility, creativity, and collaboration surged to the forefront, replacing the rigid hierarchies and standardized production methods that once defined professional environments. Work is increasingly becoming a thing you do, not a place you go to.

This paradigm shift has profound implications for the future of work. In the Information Age, the barriers to entry for creating new products, services, and ventures have dramatically lowered, paving the way for the rise of the DIY ethos in the professional realm. Individuals and small teams can now compete on a global stage, leveraging digital tools and platforms to bring innovative ideas to life.

The principles of the New Work movement align closely with these developments. This approach emphasizes the importance of meaningful work, the autonomy of workers, and the cultivation of environments that nurture innovation and creativity. New Work posits that work should not only be about economic survival but also about personal fulfillment and contributing positively to society.

At the heart of the transition from the Industrial to the Information Age is the reimagining of the workplace itself. The modern workspace has become more than just an office; it's a flexible, dynamic environment designed to foster collaboration, creativity, and the seamless integration of technology. DIY factories and coworking spaces exemplify this change, merging digital fabrication tools with community-oriented work cultures.

The adoption of technology has been instrumental in facilitating this evolution. From cloud computing and big data analysis to artificial intelligence and machine learning, these technological advancements have not only enhanced productivity but also expanded the realm of possibilities for what businesses and individuals can achieve.

However, navigating this shift requires more than just technological fluency. It demands a rethinking of traditional education and skill development, ensuring that workers are equipped to thrive in a world where automation and digital tools redefine the parameters of work. This includes fostering a mindset of continuous learning, adapting to new tools and methodologies, and cultivating a broad set of soft skills, such as critical thinking, emotional intelligence, and collaboration.

A significant consequence of the shift towards the Information Age is the blurring of boundaries between work and life. The flexibility offered by digital tools enables more and more professionals to integrate their work with their personal passions, paving the way for a more holistically fulfilling existence. Yet, this also presents challenges around work-life balance and the potential for burnout, demanding innovative solutions to ensure that the new ways of working serve to enhance, not detract from, overall well-being.

Moreover, the rise of the gig economy and freelance work reflects the changing dynamics of employment in the Information Age. These trends offer greater autonomy and flexibility for workers but also raise important questions about job security, benefits, and the social safety net. In response, new models of collective organization and social entrepreneurship are emerging to address these challenges.

Environmental sustainability has also become a critical concern in this new era. The digital age demands a reevaluation of production and consumption patterns, driving the search for sustainable practices that leverage technology to reduce waste and conserve resources. DIY factories and the maker movement exemplify this shift, emphasizing local production, recycling, and innovative uses of materials.

The global nature of the Information Age has transformed the competitive landscape, making it more accessible for businesses and individuals to participate in markets worldwide. This interconnectedness also means that trends, innovations, and crises can have far-reaching impacts, underlining the importance of global collaboration and problem-solving.

As we delve deeper into the 21st century, the transition from the Industrial to the Information Age continues to unfold, bringing new challenges and opportunities. The future of work is not fixed; it's a vibrant, evolving landscape, shaped by our collective actions, values, and vision. Embracing the principles of New Work and the DIY factory ethos offers a promising path forward—one where work is not only sustainable and innovative but also deeply fulfilling.

In conclusion, the journey from the Industrial to the Information Age is more than a narrative of technological innovation. It's a story about the potential for human growth, creativity, and collaboration. By harnessing the possibilities of this new era, we can reimagine work in ways that enrich our lives and contribute to a more equitable, sustainable, and prosperous world for all.

Principles of the New Work Movement

The New Work Movement is more than just a set of strategies; it embodies a profound shift in our approach to professional life. At its core lies a series of principles that challenge conventional norms and aspire to create a more fulfilling, innovative, and equitable work environment. These principles are the bedrock upon which the movement is built, guiding enthusiasts and professionals alike as they navigate the evolving landscape of work.

Firstly, autonomy is paramount in the New Work movement. Autonomy empowers individuals, granting them control over their workflow and decision-making processes. This principle supports the notion that creativity and innovation flourish when individuals have the freedom to explore and execute their ideas in ways that best suit their unique skills and perspectives.

Collaboration holds a central place in the movement, emphasizing the power of collective intelligence. Gone are the days of siloed departments and solitary geniuses. The New Work movement fosters environments where sharing knowledge and skills across disciplines is not just encouraged but integral. By leveraging diverse expertise, collaborative teams can solve complex problems more creatively and efficiently.

Purpose and passion drive the New Work ethos, marking a departure from the view of work as merely a means to an end. Professionals are encouraged to seek out roles that align with their personal values and aspirations, fostering a sense of fulfillment and motivation that transcends monetary compensation. This principle asserts that when people do work they truly care about, they are more engaged, productive, and likely to produce innovative outcomes.

Flexibility in work arrangements is another cornerstone of the New Work movement. This principle reflects the belief that rigid schedules and environments stifle creativity and personal well-being. By allowing for varied work hours and locations, organizations can accommodate individual needs and lifestyles, leading to a healthier work-life balance and, consequently, more dedicated and content employees.

Continuous learning and adaptability are essential in the rapidly changing landscape of work. The New Work movement encourages an environment of perpetual skill development and open-mindedness towards new methods and technologies. This commitment to growth ensures that teams remain at the forefront of innovation, prepared to pivot as necessary in response to new challenges and opportunities.

Transparency within organizations is also a key principle, promoting honesty and openness about decisions, successes, and failures. This transparency builds trust and encourages a culture of collective responsibility and learning. By making information accessible to all levels of an organization, employees become more engaged and committed to the collective vision and goals.

Equality and inclusivity are non-negotiable in the New Work movement. Recognizing the strength found in diversity, this principle advocates for the dismantling of barriers and the promotion of equitable opportunities for all, regardless of background, identity, or ability. An environment where everyone can contribute and succeed is not only more just but also more innovative and resilient.

Finally, the New Work movement emphasizes sustainability and social responsibility, urging organizations to consider the broader impact of their operations and innovations. This principle challenges businesses to go beyond profit, to contribute positively to communities and the environment, reflecting a holistic view of success that includes societal well-being.

These principles form a foundation that supports a more meaningful, innovative, and equitable work culture. As we move forward, it's crucial for forward-thinking professionals, entrepreneurs, and enthusiasts to not only understand these core values but to actively incorporate them into their work environments. By doing so, we can collectively steer towards a future where work is not just a means to an end but a source of personal fulfillment and social progress.

Embracing the New Work movement demands courage and imagination. It requires us to rethink long-held beliefs about productivity, management, and success. Yet, the potential rewards —

for individuals, organizations, and society at large — are immense. We stand on the cusp of a new era of work, one that celebrates creativity, fosters innovation, and enriches our lives in untold ways. The path forward is clear; it's up to us to take the first steps.

In conclusion, the principles of the New Work movement are not merely aspirational; they are practical guideposts that can lead us to a more innovative, fulfilling, and equitable work environment. As we delve deeper into the complexities and opportunities of the modern work landscape, let these principles light the way. Together, we can redefine what it means to work, transforming our offices, factories, and studios into spaces of inspiration, collaboration, and purpose.

Chapter 3:
The Synergy of DIY Factory and New Work

The heart of our journey into the future of work pulsates within the concept of merging DIY factories with New Work philosophies. This chapter untangles the complex, yet profoundly rewarding relationship between the hands-on, innovative mindset of the DIY movement and the flexible, technology-driven framework of New Work. Imagine a world where creativity isn't just encouraged but is the backbone of productivity; where every individual has the tools, both figuratively and literally, to carve out their unique contributions to their work environment. The DIY factory becomes a playground for the ingenious, a place where new ideas are not only born but crafted with one's own hands, reflecting the very essence of personal fulfillment and innovation.

In this innovative nexus, we see not just a convergence of ideas but a catapult that launches professionals into a realm where work is no longer a chore but a canvas. By weaving together case studies and real-world applications, we'll explore how this synergy not only addresses current workplace challenges but also paves the way for solutions that are as dynamic as they are diverse. But it's not all smooth sailing; the path is strewn with obstacles from traditional work paradigms, resistance to change, and the daunting task of harmonizing creativity with the rigors of productivity. Yet, the promise of a work culture that aligns with our deepest passions and the unyielding human spirit of innovation beckons. It's here, in this chapter, we chart the course through these challenges, drawing upon the successes and learning from the roadblocks, to mold an environment where the DIY factory and New Work aren't just concepts, but the cornerstones of a flourishing and vibrant professional landscape.

Merging Creativity with Productivity

In an era where the conventional boundaries of work are continuously being redrawn, the fusion of creativity with productivity stands as a cornerstone in the evolving relationship between DIY factories and New Work principles. This juncture represents more than just a blending of ideas; it signifies a transformative approach to how work is conceived, executed, and valued. By embedding the ethos of experimentation, innovation, and self-direction from the DIY culture into the fabric of New Work, professionals are empowered to redefine productivity beyond traditional metrics. This synergy not only elevates the quality and meaningfulness of output but also fosters a culture where creativity is not an optional add-on but a fundamental driver of progress. The dynamic interplay between these domains encourages a holistic view of work, where the pursuit of passion and the quest for efficiency coalesce to craft a future where every individual has the agency to shape their professional journey in uniquely fulfilling ways.

Case Studies: Success Stories Transitioning into the heart of innovative workplaces, there's much to be learned from those who have ventured boldly into the synergy of DIY factories and New Work. These stories aren't just tales; they're blueprints for revolutionizing the work environment. The following case studies offer a glimpse into the potential that lies within this dynamic collaboration.

The first narrative follows a small software development firm that challenged the conventional office layout and operational protocols. By adopting a DIY factory ethos, they dismantled traditional workstations and replaced them with modular, collaborative spaces.

This shift not only fostered innovation but also significantly improved team cohesion and productivity. The key takeaway? Environment shapes behavior, and by crafting spaces that encourage collaboration, workplaces can unearth new levels of creativity and efficiency.

A second notable success story comes from a manufacturing startup that fully embraced the principles of New Work by implementing flat hierarchies and inclusive decision-making processes. This approach empowered employees, making them feel truly invested in the company's goals. Consequently, the startup saw a drastic reduction in turnover rates and a surge in innovative solutions to production challenges. This example underscores the importance of reshaping organizational culture to match the evolving expectations of the workforce.

The narrative of a large corporation's transition to a DIY factory model illustrates that even the most traditional entities can adapt and thrive. By setting up internal 'innovation labs' that mimic the flexibility and creativity of DIY spaces, this corporation rejuvenated its R&D department, leading to breakthrough products. This case study illustrates how incorporating elements of the DIY ethos can renew and invigorate established companies.

Another enlightening example is a community makerspace that evolved into a fully-fledged DIY factory, serving as both a local incubator for startups and a hub for community education. This transformation highlights the potential for DIY factories not only to foster economic growth but also to serve as catalysts for community development.

In the context of New Work, a digital marketing agency redefined success by measuring outcomes not hours, transitioning to a results-only work environment (ROWE). This shift honored the diverse workstyles and life commitments of its team, leading to unprecedented levels of job satisfaction and loyalty, and, paradoxically, higher productivity and quality of work.

A publishing house, facing the industry's digital upheaval, turned to the principles of New Work and the DIY ethos to reinvent itself. By implementing collaborative editing and open-source writing platforms, they not only survived the digital transition but also became a leading voice in the open-access movement. This case stands as a testament to the power of innovation in the face of disruption.

Another compelling example is a tech company that decentralized its operations, allowing its workforce to operate remotely while still maintaining a DIY factory ethos through virtual collaboration tools. This experiment in remote work predated its widespread adoption and demonstrated the feasibility and benefits of dispersed yet tightly coordinated teams.

Furthermore, an environmental consultancy adopted the DIY factory model not just in its operational structure but also in its approach to solving ecological problems. By fostering a culture of experimentation and hands-on problem-solving, it developed groundbreaking solutions for waste management. This story reflects how the principles of DIY factories can extend beyond the realms of manufacturing and technology into other vital areas of societal concern.

Within the education sector, a university department transformed its curriculum around the DIY factory and New Work principles, preparing students not just academically but also for the realities of the modern workplace. This progressive approach has resulted in highly employable graduates capable of thinking outside conventional paradigms and equipped to navigate the challenges of tomorrow's job market.

The journey of a textile company that reinvigorated a declining industry niche by integrating New Work practices and sustainable DIY production techniques underscores the revitalizing potential of these concepts across various sectors. Their success in marrying tradition with innovation serves as a compelling case for the adaptability and relevance of these principles.

A unique story emerges from a healthcare startup that applied New Work principles to address burnout and inefficiency among medical professionals. By redesigning workflows and encouraging autonomy, they not only enhanced patient care but also improved the work-life balance of their staff, illustrating that New Work principles hold transformative potential even in the highly structured domain of healthcare.

In a striking example of cross-industry innovation, an automotive company borrowed from the playbook of DIY factories to set up collaborative workspaces where engineers and designers from different specializations and backgrounds could come together to solve complex problems. This initiative led to the development of groundbreaking vehicle designs and technologies, proving the immense value of interdisciplinary collaboration.

An entertainment company, struggling to keep pace with the rapid changes in consumer behavior and technology, turned its fortunes around by adopting a DIY ethos. By empowering creative teams to experiment freely and fail fast, they launched a series of successful products that captured the imagination of a global audience. Their turnaround story emphasizes the significance of fostering a culture of innovation and resilience.

The transformation of a traditional retail chain into an experiential space that combines shopping with DIY workshops and innovation labs speaks volumes about the versatility and appeal of the DIY factory concept. This reimagining of retail has not only revived the brand but also created a new model for engaging and inspiring customers in a digital age.

Lastly, an international development organization incorporated DIY factory and New Work principles to enhance collaboration and innovation in solving global challenges. By breaking down silos and encouraging cross-functional teams to work on projects with a high degree of autonomy, they significantly improved the effectiveness and impact of their initiatives. This example showcases the universal applicative potential of these forward-thinking concepts.

The narratives detailed above provide a litany of evidence that the fusion of DIY factories and New Work principles isn't just a fleeting trend—it's a robust framework for fostering innovation, productivity, and meaningful work across a myriad of contexts. Embracing these practices can lead organizations, regardless of size or sector, into a prosperous future characterized by creativity, collaboration, and an unwavering commitment to improvement.

Challenges and Solutions

Merging the ethos of the DIY factory with the principles of New Work presents a unique set of challenges. These obstacles can seem daunting, but within each challenge lies the seed of innovation and opportunity. Let's explore some of these challenges and the creative solutions that can turn them into stepping stones for progress.

One of the first challenges is the adjustment to a shift in mindset. Traditional work environments are structured around hierarchy and specialization, whereas the New Work model emphasizes flat structures and multi-disciplinary collaboration. To bridge this gap, organizations can initiate workshops and team-building activities that foster a sense of community and shared purpose. These efforts help in cultivating an environment where creativity and autonomy are not just encouraged but are seen as essential to the workflow.

Another significant challenge lies in resource allocation. DIY factories require access to tools, technologies, and spaces that facilitate experimentation and innovation. This can be a hurdle for startups or smaller enterprises with limited budgets. The solution? Embrace the power of community. By collaborating with maker spaces, sharing resources with other small businesses, or even initiating crowd-funding campaigns, organizations can overcome financial constraints and create rich, resourceful ecosystems for innovation.

The integration of technology also poses a challenge. The rapid pace of technological advancement means that staying up-to-date requires both time and investment. The solution here is twofold. First, create a culture of continuous learning within your organization, where acquiring new skills and knowledge is part of everyday work life. Second, foster partnerships with tech companies and academic institutions. These relationships can provide access to cutting-edge technologies and expertise, creating a symbiotic environment where both parties benefit from shared innovation.

Scaling up while maintaining the ethos of DIY and New Work is another challenge. As organizations grow, there's a risk of losing the close-knit community feel and the fluidity that makes these models so dynamic. To address this, it's crucial to implement systems that support scalability without compromising on core values. This can include developing clear but flexible guidelines, investing in communication tools that foster transparency and connection, and ensuring that all members of the organization are aligned with its mission.

Furthermore, the traditional performance metrics used in many industries do not adequately capture the value created by the synergy of DIY factories and New Work. This calls for a new approach to measuring success, one that factors in innovation, employee engagement, community impact, and environmental sustainability. Developing these metrics can be a collaborative effort that involves input from all levels of the organization, ensuring that they reflect the collective vision and goals.

Legal and regulatory issues can also arise, particularly when it comes to intellectual property and labor laws. Navigating this complex landscape requires a proactive approach. Engaging with legal experts to understand the implications of collaborative work and shared innovation is essential. Additionally, advocating for policies that support the New Work model can help create a more conducive environment for these practices to flourish.

Attracting and retaining talent is a challenge that many organizations face, but it's particularly pertinent for those embracing the DIY and New Work models. These models demand a high degree of autonomy, creativity, and adaptability – qualities that are not nurtured by traditional education systems. To attract the right talent, organizations need to communicate their vision effectively and offer opportunities for growth and learning that go beyond conventional career trajectories. Moreover, creating a supportive and inspiring work environment can help retain this talent.

Finally, fostering a culture of experimentation and risk-taking can be challenging in environments where failure is seen negatively. To overcome this, organizations should celebrate both successes and failures – viewing every outcome as a learning opportunity. Encouraging small-scale experiments and adopting a flexible approach to projects can help build a culture where innovation thrives.

In conclusion, while the challenges of integrating DIY factory principles and New Work concepts are real and varied, they are not insurmountable. By adopting creative solutions that focus on collaboration, community, and continuous learning, organizations can navigate these challenges successfully. The journey toward a more innovative, inclusive, and fulfilling work environment is a collaborative effort – one that requires patience, resilience, and an unwavering belief in the potential of synergy between these two powerful movements.

Chapter 4:
Setting the Stage for Innovation

In a world where the only constant is change, "Setting the Stage for Innovation" serves as a crucial blueprint for those eager to spearhead a renaissance in modern work culture. At the heart of any transformative movement is an environment that not only fosters creativity but actively encourages the collision of ideas and disciplines. As we delve into the nuances of designing collaborative workspaces, we uncover the alchemy of physical and digital realms that propels the DIY factory concept into uncharted territories of efficiency and inventiveness. It's not just about selecting the right tools and technologies but about crafting ecosystems that nurture a community of innovators, where the exchange of knowledge ignites the spark of breakthroughs. This chapter is dedicated to empowering forward-thinking professionals, entrepreneurs, and enthusiasts with the insights needed to architect the foundations of such innovative environments. Through strategic design, embracing diversity, and fostering a spirit of collaboration, we pave the way for the DIY factories and New Work principles to thrive, setting the stage for a future where work is not just a means to an end but a continuous journey of exploration, growth, and transformation.

Designing Collaborative Workspaces

In envisioning the future of work, the creation of collaborative workspaces stands out as a critical pivot towards fostering innovation.

Such environments are meticulously designed to blur the lines between work and creativity, encouraging spontaneous encounters and the free flow of ideas among individuals. Imagine a space where the physical layout itself acts as a catalyst for teamwork, with open-plan areas that invite interaction, private nooks for deep thought, and adaptable work stations that fit the task at hand. It's not just about placing a few desks together or adding a ping-pong table; it's about creating an ecosystem that supports the dynamic nature of innovative work. Integrating technology seamlessly, from cloud-based collaboration tools to state-of-the-art hardware, ensures that these workspaces are not only conducive to current projects but are adaptable for future challenges as well. In designing these spaces, one must consider how every element, from lighting to furniture, can contribute to an atmosphere of collective effort and boundless potential. This endeavor is not just about physical space but about building the foundation of a culture that values openness, flexibility, and mutual respect. The goal is clear: to craft environments where every professional, entrepreneur, and enthusiast can thrive, driving the DIY ethos and new work concepts forward in a world that craves innovative solutions.

Tools and Technologies for the DIY Factory In the pursuit of merging the DIY ethos with the modern workspace, it's vital to understand the tools and technologies that can turn this vision into reality. Embracing the ideology that drives the DIY factory necessitates an exploration into the innovative, sometimes disruptive, technologies that enable individuals and teams to create, innovate, and produce in ways that were unthinkable just a few decades ago.

The foundation of any DIY factory lies in its ability to harness the potential of digital fabrication tools. 3D printers, for example, have transcended their initial novelty status to become essential tools for rapid prototyping, custom manufacturing, and even large-scale production endeavors. The accessibility of these printers has empowered creators at all levels, allowing for the materialization of ideas with a speed and precision that rivals traditional manufacturing processes.

Laser cutters and CNC machines similarly democratize the ability to work with a variety of materials, offering precision cutting, engraving, and milling capabilities. These tools, once exclusively the domain of industrial factories, are now finding their place in workshops of all sizes, enabling the production of complex designs that would be challenging to achieve by hand.

A crucial element that synergizes with these physical tools is software. CAD (Computer-Aided Design) and CAM (Computer-Aided Manufacturing) software have become more user-friendly, with communities around the world developing open-source options that are available to anyone with a computer. This software enables individuals to design intricate parts, plan their projects with precision, and directly control digital fabrication machines, essentially bridging the gap between idea and object.

The integration of IoT (Internet of Things) devices into the DIY factory space opens a new frontier for innovation. Smart sensors can monitor environmental conditions, machine performance, and even track inventory in real-time, optimizing the manufacturing process and reducing waste. This level of interconnectedness not only boosts productivity but also introduces a new layer of creativity in how projects are managed and executed.

Robotics, including affordable robotic arms and mobile robots, are increasingly becoming part of the DIY landscape. These robots extend the capabilities of human makers, offering assistance in repetitive or precision tasks, and opening up new possibilities for complex, collaborative projects between humans and machines.

Virtual and augmented reality technologies are also finding their place in the DIY factory. From virtual prototyping to augmented assembly instructions overlaid onto the real world, these technologies offer innovative ways to design, learn, and collaborate. They significantly reduce the time and resources needed to go from concept to creation, further blurring the lines between digital and tangible realms.

Behind these tools and technologies lies a powerful framework for collaboration and learning. Online platforms and communities play a critical role in the DIY factory paradigm, offering spaces for sharing knowledge, sourcing inspiration, and crowd-sourcing solutions to complex challenges. These platforms not only provide access to a vast repository of information but also connect like-minded individuals, fostering a culture of collective innovation and support.

The potential of AI and machine learning in this context cannot be understated. From optimizing designs for sustainability and functionality to predicting failures before they happen, AI can offer insights and efficiencies that human makers might miss. This intelligent layer of technology, integrated thoughtfully, can enhance creativity rather than replace it, by automating the mundane and illuminating new paths for innovation.

However, adopting these tools and technologies is not merely about acquisition but about mindset. A culture that encourages experimentation, learning from failure, and sharing knowledge is crucial. It's about recognizing that these technologies are not just tools but co-creators in the DIY factory, each with the potential to open new avenues for expression, innovation, and production.

Furthermore, sustainability is a vital consideration in the selection and use of these tools. Energy-efficient machines, recycled materials for 3D printing, and sustainable supply chains are not just ethical choices but also align with the efficiency and innovation goals of the DIY factory. The technology chosen should enable a reduction in waste, an increase in recycling, and a move towards more sustainable production methods.

The evolution of these tools and technologies is ongoing, with advances happening rapidly. Staying informed and adaptable, ready to integrate new developments into the DIY factory, is part of this dynamic landscape. It's a journey of continuous learning, where the next breakthrough is always just around the corner, ready to transform what's possible in the world of DIY manufacturing.

Ultimately, the tools and technologies for the DIY factory are about empowering individuals and teams to turn their visions into reality, without the traditional barriers of cost, access, or expertise. They represent not just a shift in how we produce but in how we conceive of production itself - as a creative, collaborative, and accessible endeavor.

The journey to integrating these tools and technologies into your workspace is not just about setting up equipment. It's about building a culture of innovation, creativity, and shared purpose. It's about envisioning a future where ideas flow freely, obstacles are opportunities for learning, and the boundary between imagination and reality is ever more porous. The DIY factory, equipped with these tools and driven by these principles, is a powerful blueprint for the future of work.

As we look towards that future, it's clear that the potential is not just in the technologies themselves but in how we choose to use them. It's in the projects that challenge us, the problems we solve for our communities, and the world we envision and build together. The tools and technologies for the DIY factory are at our fingertips, waiting for us to grasp them and create the future we imagine.

Fostering a Community of Innovators

When we embark on the journey of nurturing a culture steeped in innovation, we're not just talking about erecting a physical space filled with the latest gadgets and technology. We're delving into the creation of a vibrant ecosystem where every individual feels empowered, inspired, and connected to a larger purpose. This is where the true magic happens, and it's the bedrock upon which DIY factories and New Work principles thrive.

Central to this ecosystem is the concept of community. Not just any community, but one that's dynamically woven together by shared values, visions, and the relentless pursuit of pushing boundaries. It's an environment that beckons the curious, the dreamers, the doers, and the relentless innovators to come together and share their unique perspectives.

However, forging such a community doesn't occur by chance. It requires intentional design and thoughtful cultivation. One of the first steps is to actively seek diversity in skills, experience, and thinking. Innovation thrives on varied perspectives. When people from different backgrounds and disciplines collide, they bring with them unique insights that can spark unexpected solutions and breakthrough ideas.

Communication is the lifeblood of any thriving community. Establishing open channels where ideas, successes, and even failures can be shared without judgment fosters a culture of trust and mutual respect. This openness not only drives collaboration but also ensures that learning is a continual process. Each setback or success becomes a lesson for the collective, propelling everyone forward together.

Equally important is creating a sense of ownership and accountability. When individuals feel that they're an integral part of the community and that their contributions matter, they're more likely to invest their time, effort, and creativity. This investment doesn't just fuel their personal growth but accelerates the collective momentum towards innovation.

To stimulate creativity and innovation, the physical and virtual spaces we create play a pivotal role. They should inspire and facilitate the free flow of ideas. Think open workspaces that are configurable for different needs, communal areas designed for spontaneous interactions, and digital platforms that allow for collaboration across boundaries. The environment should be as adaptable as the ideas it aims to nurture.

But fostering a community of innovators isn't only about creating the right physical and digital spaces. It's also about crafting and maintaining the right atmosphere. This involves celebrating risks and valuing the learning derived from trying something new, even when it doesn't work out as planned. It's about championing the process of innovation, not just the outcomes.

Leadership within such communities also takes on a new meaning. Instead of traditional top-down directives, leaders serve as catalysts and facilitators — they're there to guide, inspire, and remove obstacles from the path of innovation. Their role is to nurture and protect the culture, ensuring that the community remains resilient, dynamic, and inclusive.

Encouraging cross-pollination of ideas is another cornerstone. Organize hackathons, workshops, and speaker events that not only bring the community together but also introduce fresh perspectives and new challenges. These interactions can serve as the spark for innovative ideas that might not have emerged in isolation.

Within this vibrant community, it's critical to also have mechanisms for recognizing and rewarding innovation. This recognition shouldn't be limited to successful outcomes but should also celebrate the courage to experiment and the insights gained from the process. Such rewards further reinforce the values of creativity, experimentation, and continuous learning.

As the community matures, it's vital to remain flexible and adaptable. The needs, challenges, and opportunities will evolve, and so must the community. Staying attuned to the pulse of the community and being willing to pivot or transform aspects of the ecosystem ensures its long-term vitality and relevance.

Participation in the larger ecosystem outside of the immediate community can also stimulate innovation. Engaging with other innovators, industries, and sectors provides new stimuli and opportunities for collaboration. It's through these connections that the community can truly amplify its impact and contribute to wider societal and economic development.

At its core, fostering a community of innovators is about cultivating a space where everyone is encouraged to question, explore, and imagine. It's about creating a culture that not only tolerates ambiguity and uncertainty but celebrates the creative possibilities they bring.

Finally, remember that building such a community is a journey, not a destination. It requires patience, commitment, and a shared belief in the power of collaboration and innovation. But the rewards — a thriving ecosystem of innovators poised to make meaningful contributions to the world — are well worth the effort.

As we continue to explore the principles of New Work and the dynamics of DIY factories, let's carry forward the understanding that at the heart of any technological or work revolution lies a deeply human element. It's the spirit of community and collaboration that will truly drive us into a future brimming with possibility and innovation.

Chapter 5:
The DIY Factory Model

In essence, the DIY Factory Model reimagines the production landscape, blending the ethos of the do-it-yourself movement with cutting-edge innovation to create spaces where creativity and productivity intersect in vibrant, dynamic ways. This model doesn't just challenge the conventional boundaries of workspaces; it dismantles them, encouraging a fluid, adaptable approach to manufacturing and design. By integrating key components like open-source technologies, collaborative work areas, and sustainability practices, DIY factories stand as a testament to the potential for workplaces to be not just centers of economic activity, but hubs of community and creativity. Operating a DIY factory requires a mindset shift—from a top-down management style to a more decentralized, participatory approach that values every team member's contributions. This democratization of the production process does more than boost morale; it leads to innovation and efficiency improvements by leveraging a diverse range of perspectives and skills. As we traverse this chapter, we aim to unpack the anatomy of a successful DIY factory, delve into its operation, and explore the tangible benefits it brings to the workforce and beyond. In doing so, we invite you to envision a future where the landscape of work is reshaped by the principles of collaboration, innovation, and a steadfast commitment to sustainability.

Anatomy of a DIY Factory

At the heart of the DIY Factory model lies a beacon of innovation, a framework that not only breaks the mold of traditional manufacturing but also seamlessly integrates with the principles of New Work. It's a space where creativity meets productivity, and where the machinery of creation is as diverse as the minds that operate them. The anatomy of a DIY Factory is characterized by its dynamic components and layout, designed to foster collaboration, innovation, and a relentless pursuit of efficiency. Imagine a floor filled with 3D printers, laser cutters, and CNC machines, all buzzing in the background as teams huddle over blueprints and prototypes. Here, the assembly line is not just a path products travel but a journey of ideas coming to life, evolving with each iteration. This model champions adaptability, pushing us to think beyond fixed roles and traditional workflows, encouraging a mindset where challenges are met with creative solutions. It's about embracing the power of community, where knowledge and skills are shared freely, lighting the path for others to follow. As we delve deeper into the DIY Factory's anatomy, we uncover the essence of modern manufacturing, a testament to human ingenuity and the relentless drive to create, innovate, and inspire a new era of work.

Key Components and Layout Continuing from our exploration into the anatomy of a DIY factory, it's imperative to dive into the core aspects and physical layout that are quintessential for catalyzing innovation and fostering a culture of creativity. The layout of a DIY factory isn't just about placing equipment and workstations; it's about crafting an ecosystem that nurtures collaboration, experimentation, and continuous learning.

The heart of a DIY factory lies in its ability to be adaptable. The layout must be modular, allowing areas to be reconfigured as projects evolve and teams' needs change. This flexibility supports the iterative nature of creative work, where projects can shift directions based on new insights or feedback. Imagine spaces that transform effortlessly from woodworking workshops to electronics labs, embodying the spirit of innovation that drives the DIY ethos.

A key component of this layout is the communal areas designed to facilitate unplanned interactions and serendipitous collaborations. These spaces are the breeding grounds for new ideas, where professionals from diverse backgrounds can share insights and spark off each other's creativity. Think of large, open-plan areas dotted with comfortable seating, coffee stations, and whiteboards brimming with sketches and mind maps.

Another crucial element is the integration of state-of-the-art tools and technologies. From 3D printers and laser cutters to advanced software for design and prototyping, a DIY factory must equip its inhabitants with the tools to turn their visions into tangible realities. However, it's not just about having the right tools; it's about making them accessible to all, demystifying technology, and empowering every member to bring their ideas to life.

Equally important is the creation of dedicated zones for concentration and deep work. While collaboration is key, innovation also requires periods of individual focus and uninterrupted flow. These areas should be sanctuaries for thought, where noise and distractions are minimized, allowing individuals to delve deep into the complexities of their work.

Learning and skill-sharing play pivotal roles in the culture of a DIY factory. Classrooms, workshop areas, and even online platforms should be incorporated into the layout to facilitate ongoing education and mastery of new skills. Positioned as accessible learning hubs, they encourage a culture of continuous improvement and open the door to cross-disciplinary exploration.

The incorporation of green spaces and elements of nature within the factory layout isn't just an aesthetic choice; it's a strategic one. Research suggests that proximity to natural elements boosts creativity, reduces stress, and enhances well-being. Whether it's indoor plants, green walls, or outdoor relaxation areas, integrating nature into the workspace can have a profound impact on innovation and productivity.

Sustainability should be woven into the fabric of a DIY factory, not just in operational processes, but in the physical layout as well. Utilizing renewable energy sources, incorporating materials recycling stations, and designing for energy efficiency are all ways the layout can reflect a commitment to environmental sustainability.

Storage and material management are often overlooked but are critical for maintaining the functionality and efficiency of a DIY factory. Thoughtful storage solutions that are both accessible and organized, ensure that materials are readily available, but also that the workspace remains uncluttered and conducive to creativity.

Visibility and transparency within the workspace encourage accountability, inspire participation, and foster a culture of open feedback and collaboration. Designing work areas so projects are visible to all promotes a shared sense of purpose and achievement, reinforcing the community ethos that is central to the DIY factory model.

Safety is paramount in any workspace, especially one like a DIY factory, where varied and potentially hazardous tools and machinery are in regular use. The layout must prioritize safety, with clear signage, easily accessible safety equipment, and defined pathways that minimize risks without stifacing creativity and productivity.

Community areas that go beyond work, such as communal kitchens, lounges, and perhaps even gaming or relaxation zones, support well-being and job satisfaction. These spaces provide opportunities for team members to connect on a personal level, building the kind of trust and camaraderie that's essential for true collaboration.

The digital infrastructure of a DIY factory must be seamlessly integrated into its physical layout. High-speed internet, wireless connectivity, and digital collaboration tools should be omnipresent, supporting a blend of physical and virtual collaboration that's increasingly relevant in today's work environment.

The exterior design and accessibility of the DIY factory also matter. It should not only inspire those who work within its walls but also beckon the community outside. A visually appealing, accessible entrance, possibly with public showcases or galleries, can bridge the gap between the factory and the wider community, inviting public engagement and fostering a broader culture of innovation.

In conclusion, the key components and layout of a DIY factory are about far more than the physical space. They're about creating an environment that lives and breathes innovation, collaboration, and sustainability. Every aspect of the design must serve to empower individuals and teams to explore, create, and transform their ideas into reality. This isn't just about building a workspace; it's about fostering a vibrant community of thinkers, makers, and innovators who are equipped to lead the charge in the new era of work.

Operating a DIY Factory

In the heart of any DIY Factory lies the imperative to seamlessly blend the ethos of do-it-yourself culture with the rigors of productive enterprise. Operating such a facility goes beyond merely providing tools and space; it's about cultivating a mindset of continuous innovation, collaboration, and self-sufficiency. Imagine a place where every individual is empowered to experiment, failures are seen as stepping stones to mastery, and the boundary between work and creative expression is deliberately blurred. Managers in a DIY Factory wear multiple hats - they're facilitators, mentors, and sometimes, learners themselves. They foster an environment where sustainable practices are not just encouraged but are the norm, ensuring that the factory not only thrives in its outputs but does so responsibly. This operational model champions the belief that with the right structure, community, and ethics, manufacturing can be revolutionized to be more inclusive, innovative, and impactful. Through lean operational frameworks and adaptive leadership, the DIY Factory isn't just a production facility; it's a crucible for the new work movement, setting a precedent for future enterprises aiming to merge creativity with productivity in a sustainable and empowering manner.

Management and Sustainability In the realm of DIY factories and the ever-evolving landscape of New Work, management and sustainability are not merely buzzwords but foundational principles that drive long-term success and innovation. As we traverse this chapter, we dive deep into how forward-thinking professionals, entrepreneurs, and enthusiasts can mold these principles into their operational and cultural fabric, ensuring their ventures not only thrive but become beacons of sustainable innovation in the modern work era.

The essence of managing a DIY factory intertwines closely with the ethos of sustainability. This holistic approach extends beyond environmental aspects, embedding itself into economic models, community engagement, and the well-being of all participants. Embracing sustainability means considering the life cycle of every project and product, from conception through to its end-of-life, ensuring resources are used efficiently, and waste is minimized.

Leadership within these spaces must champion a culture of continuous learning and adaptation. The pace at which technology and markets evolve demands an agile approach to management. It involves empowering teams, fostering an environment where feedback is not only encouraged but acted upon, and where failure is seen as a stepping stone to innovation rather than a setback.

At the heart of sustainable management practices within DIY factories lies the commitment to open, flat hierarchies. This management style not only promotes transparency but also encourages a culture of shared responsibility. It enables faster decision-making, leverages diverse viewpoints, and aligns with the DIY ethic of collaboration and collective creativity.

Financial sustainability is a critical component, driving the need for innovative funding models. Beyond traditional sources like venture capital, embracing crowdfunding, community investments, or revenue-sharing models can provide the financial backbone for growth while ensuring the community feels invested in the venture's success.

Operational sustainability also plays a pivotal role. Adopting lean manufacturing principles, even in the context of a DIY factory, can lead to significant efficiencies. This might include minimizing resource usage, optimizing production processes, or adopting just-in-time inventory systems to reduce waste.

Environmental responsibility must be woven into the management fabric. This involves adopting renewable energy sources, ensuring materials are sustainably sourced, and implementing recycling programs. Such practices not only reduce the carbon footprint but also set a benchmark in the industry for environmental stewardship.

In the sphere of New Work, the human element of sustainability is paramount. Creating a work environment that prioritizes mental and physical health, encourages work-life balance, and supports personal growth is essential. This fosters a motivated, innovative, and resilient workforce prepared to tackle the challenges of the future.

Community engagement and social value generation are inseparable from the management structures of DIY Factories. By engaging with local communities, educational institutions, and other stakeholders, ventures can build robust ecosystems that support mutual growth, skills development, and social cohesion.

Technological sustainability involves staying abreast of technological advancements while also weighing their implications on society, the environment, and the economy. It means selecting technologies that not only enhance productivity and creativity but are also accessible, ensuring broad participation and minimizing technological divides.

Data management and privacy practices reflect another crucial management consideration in today's digital age. Adhering to best practices in data protection not only builds trust but ensures compliance with an ever-evolving regulatory landscape, protecting both the enterprise and its stakeholders.

Measurement and reporting of sustainability outcomes enable DIY factories to not just aim for success but to quantify it. This goes beyond traditional financial metrics to include social, environmental, and economic impacts, providing a comprehensive view of the venture's contribution to society.

Adaptability and resilience, key tenets of the New Work movement, underscore the importance of future-proofing through sustainable management practices. This includes preparing for future challenges by fostering innovation, developing strategic partnerships, and continuously scanning the horizon for emerging trends and disruptions.

Finally, sharing knowledge and best practices plays a vital role in the management and sustainability of DIY factories. By engaging in open source projects, participating in industry forums, and collaborating with other ventures, knowledge is disseminated, enriching the entire ecosystem.

As we embark on this journey towards sustainable management within the DIY factory and New Work domains, it is evident that the challenges are myriad but not insurmountable. By adopting a holistic approach to management that encompasses economic, environmental, and social dimensions, ventures can pave the way for a future where work is not only productive and innovative but also responsible and sustainable. Let us embrace these principles, leading with purpose and passion, to craft a future that harmoniously blends creativity, innovation, and sustainability.

Chapter 6:
New Work Structures and Organizational Culture

In an age where agility and innovation set the pace for success, organizations are rapidly shifting from traditional hierarchies to more fluid structures that embrace open communication, autonomy, and continuous adaptation. This chapter embarks on a journey to unravel the very fabric of these emerging work environments, where leadership is distributed, and decision-making is democratized. We dive deep into the dynamics of flat hierarchies where everyone's voice matters, fostering a culture of trust and mutual respect. Essential to this transformation is the implementation of team autonomy – a powerful catalyst for unleashing creativity and accelerating performance. We explore strategies to empower teams, giving them the freedom to navigate challenges, innovate, and achieve their collective goals without the constant need for top-down directives.

Equally, we probe into the heart of organizational culture, advocating for a paradigm where continuous learning and adaptation are not just encouraged but embedded into the very essence of the company ethos. This culture transformation demands more than just structural changes; it requires a mindset shift at all levels of the organization. Through a meticulous blend of inspirational narratives and actionable insights, we chart a course for building resilient work cultures that not only adapt to change but thrive on it. By embracing these new work structures and cultivating an organizational culture that prizes learning, adaptation, and autonomy, businesses can navigate the uncertainties of the future with confidence and creativity.

Flat Hierarchies and Open Communication

In a world where the speed of information and the pace of change are constantly accelerating, traditional hierarchical structures in work environments can often hinder rather than help. Embracing flat hierarchies and open communication within the framework of new work structures and organizational culture isn't just a trend – it's a transformative strategy that drives innovation, fosters creativity, and empowers employees. By flattening the hierarchy, organizations can dismantle the barriers that silence voices and impede the free flow of ideas. This model prioritizes transparency, where feedback and dialogue are encouraged across all levels, creating a culture where everyone feels invested in the company's success. Open communication enhances collaboration and facilitates agile response to challenges, making the company more adaptive and resilient. As we navigate through the complexities of modern work environments, integrating flat hierarchies and open communication into the heart of organizational culture not only liberates individuals to achieve their full potential but also propels organizations towards unparalleled growth and innovation.

Implementing Team Autonomy Within the fabric of New Work and the DIY factory ethos lies a golden thread: autonomy. Not merely a buzzword, but a structured paradigm shift that empowers teams to navigate the complexities of modern projects with innovation, creativity, and swiftness. To implement team autonomy effectively, understanding its foundation and the strategies to cultivate it is paramount.

At its core, team autonomy is the freedom for teams to make decisions without micromanagement, alongside the responsibility to own the outcomes of those decisions. It fosters a sense of ownership, encourages risk-taking, and facilitates a deeper understanding of work processes among team members. However, transitioning to this model requires meticulous planning, communication, and a gradual shift in organizational culture.

The first step towards fostering autonomy is trust. Leaders must trust that their teams have the organization's best interests at heart and are capable of making sound decisions. This trust doesn't sprout overnight; it's cultivated through consistent and positive reinforcement of team decisions. Begin by allowing teams to make small decisions, gradually increasing their scope as confidence in their judgment grows.

Transparency is another cornerstone of implementing team autonomy. Transparent communication about the organization's goals, challenges, and expectations provides teams with the context needed to make informed decisions. It also demystifies the decision-making process, making it more accessible and less intimidating for team members to contribute their ideas and feedback.

Establishing clear boundaries is critical. Autonomy does not imply a free-for-all scenario. Instead, clear guidelines about what decisions teams can make and the boundaries of their autonomy prevent confusion and potential overreach. These boundaries should be flexible enough to allow creativity and innovation to flourish while keeping the team aligned with the organization's overall objectives.

Empowering teams with the right tools and resources is essential for autonomy to take root. Whether it's access to information, technology, or training, providing teams with what they need to execute their decisions is fundamental. This empowerment also includes offering support when teams take risks and fail, treating these instances as learning opportunities rather than setbacks.

Building a supportive community around autonomous teams encourages a culture of sharing, collaboration, and mutual support. Encourage teams to share their successes and challenges in implementing autonomy with the rest of the organization. This not only fosters a sense of community but also provides valuable insights that can refine the process of autonomy implementation.

Feedback loops are vital. Regular, constructive feedback helps teams understand the impact of their decisions and fosters a culture of continuous improvement. This feedback should not be top-down but also peer-to-peer, creating a comprehensive feedback ecosystem that values everyone's perspective.

Training and development play a significant role in preparing teams for autonomy. Training programs focused on decision-making, leadership, and effective communication equip team members with the skills necessary to navigate autonomy successfully. Moreover, development opportunities that align with the individuals' career goals and the organization's needs ensure that team autonomy contributes to both personal and organizational growth.

Reward and recognition systems that align with autonomous working models reinforce the behavior you want to see. Celebrating successes, big and small, acknowledges the effort and risks taken by the teams and encourages them to continue pushing boundaries.

Implementing team autonomy is not without challenges. Resistance from both management and team members accustomed to traditional hierarchies can hamper the transition. Addressing concerns openly, providing clear evidence of the benefits of autonomy, and involving skeptics in the process can mitigate resistance and foster buy-in.

Moreover, autonomy requires a shift in mindset from both leaders and team members. Leaders must learn to relinquish control and become facilitators rather than directors. At the same time, team members must step up, taking more initiative and responsibility than they might be accustomed to. This cultural shift is perhaps the most challenging aspect of implementing team autonomy but also the most rewarding.

Finally, patience is key. The transition to an autonomous team model doesn't happen overnight. It's a gradual process that evolves as the organization and its people grow. Celebrating milestones, learning from setbacks, and continuously refining the approach towards autonomy are essential steps on this journey.

In summary, implementing team autonomy within the framework of New Work and the DIY factory ethos encapsulates a vision of a more engaged, innovative, and fulfilling work environment. By building on trust, transparency, empowerment, and community, organizations can unleash the full potential of their teams, paving the way for unprecedented levels of creativity and efficiency. The future of work is not about controlling resources but unleashing human potential, and team autonomy is a beacon on this transformative journey.

Building a Culture of Continuous Learning and Adaptation

In an era where change is the only constant, organizations are facing an unprecedented need to evolve rapidly and efficiently. The intersection of the DIY ethos with new work structures offers a compelling blueprint for fostering a culture that not only embraces change but thrives on it. At the heart of this evolution is the commitment to continuous learning and adaptation.

Imagine an organization where every individual sees learning as an integral part of their job, not an optional add-on. In such an environment, employees are not just workers; they are learners, innovators, and changemakers. This is not a distant utopia but a tangible reality that organizations can achieve by embedding continuous learning into their DNA.

Continuous learning transcends traditional training programs. It's about creating a mindset where every experience, challenge, and feedback loop is an opportunity to grow. It means shifting from a focus on qualifications and titles to a focus on skills, creativity, and adaptability. This shift requires a fundamental change in how both leaders and employees view their roles in the organization.

Leadership plays a pivotal role in cultivating this culture. Leaders must lead by example, demonstrating their commitment to their personal growth and encouraging their teams to do the same. This involves not only supporting their teams in pursuing learning opportunities but also recognizing and rewarding curiosity, experimentation, and resilience in the face of failure.

Embracing a continuous learning culture also means redesigning work processes to support constant evolution. It requires flattening hierarchies to facilitate open communication, where ideas and feedback flow freely across all levels of the organization. Team autonomy is crucial here, allowing individuals and teams to test, learn, and iterate rapidly without being hamstrung by bureaucratic red tape.

Moreover, technology plays a critical role in supporting a learning-oriented workplace. From collaborative platforms that bring teams together to learn and solve problems in real-time, to AI-driven tools that offer personalized learning experiences, technology can significantly amplify learning opportunities within the organization.

However, building a culture of continuous learning is not without its challenges. It demands a willingness to disrupt the status quo, to let go of outdated practices and notions of how 'work' should be done. It requires a commitment to resilience, understanding that failure is not just possible but a valuable stepping stone in the learning journey.

This journey also involves rethinking the physical and virtual spaces where work happens. Creating environments that stimulate creativity and collaboration, whether through innovative use of physical workspace or virtual collaboration tools, is essential for fostering the kind of dynamic interaction that fuels continuous learning.

Furthermore, fostering a community of innovators within the organization amplifies the impact of a continuous learning culture. When individuals feel part of a supportive community, they are more likely to take risks, share knowledge, and collaborate on innovative solutions.

One of the most powerful aspects of building a culture of continuous learning and adaptation is its capacity to attract and retain top talent. Today's professionals are looking not just for a job, but for opportunities to grow, innovate, and make a meaningful impact. Organizations that offer such environments become talent magnets, drawing in the best and brightest.

In addition to attracting talent, a learning-oriented culture also plays a critical role in an organization's ability to adapt to external changes. Whether it's navigating technological advancements, shifting market dynamics, or responding to societal challenges, a workforce that is constantly learning is better equipped to pivot and innovate.

Implementing a culture of continuous learning and adaptation also involves recognizing and leveraging the diverse talents and perspectives within the organization. This diversity is a powerful engine for innovation, as it brings together a wide range of ideas, experiences, and approaches to problem-solving.

To truly embed continuous learning into the organizational culture, it's vital to measure and celebrate progress. This doesn't mean solely focusing on traditional performance indicators but also developing metrics that capture learning, innovation, and adaptation. Celebrating milestones and learning from every outcome fosters a positive feedback loop that encourages continued growth and exploration.

Ultimately, building a culture of continuous learning and adaptation is not a one-time initiative but an ongoing journey. It's about creating an environment where change is embraced, where challenges are met with creativity and resilience, and where every individual feels empowered to learn, innovate, and drive positive change.

In conclusion, the intersection of DIY factories and new work structures presents a compelling paradigm for organizations seeking to thrive in an ever-changing world. By embracing continuous learning and adaptation, organizations can not only navigate the complexities of the modern landscape but also shape the future of work in ways that foster innovation, engagement, and sustained growth.

Chapter 7:
The Role of Technology in
Shaping the Future of Work

The transformative power of technology in reshaping the contours of our work cannot be overstated. As we delve into the heart of what makes up the future of work, it's evident that the tools, platforms, and systems we employ are not just facilitators but active architects of new paradigms. Imagine a landscape where automation does not supplant creativity but serves as its catalyst, where AI and robotics in the DIY factory enhance human effort, allowing us to transcend traditional boundaries of productivity and innovation. This synergy between technology and human ingenuity is the cornerstone of New Work practices, enabling a fluidity and flexibility previously unimaginable. By leveraging technology, we're not just optimizing for efficiency; we're unleashing a wave of creative potential, paving the way for workspaces that are more adaptable, resilient, and ultimately more fulfilling. In this chapter, we'll explore how technology acts as an enabler, transforming not just how we work, but why we work, heralding a future where work is not a place we go, but a thing we do with passion, purpose, and profound impact.

Automating for Efficiency and Creativity

In the dynamic interplay between technology and the evolving landscape of work, automation emerges as a pivotal force, not merely streamlining efficiency but also unlocking new realms of creativity. Imagining a future where machines handle mundane tasks, we're afforded the luxury of time—time to innovate, to dream, and to sculpt our professional environments into spaces of unparalleled creative flourishing. This isn't about rendering human roles obsolete but rather redefining them. It challenges us to leverage automation in a way that complements and enhances human ingenuity, inviting a collaboration between man and machine that propels both productivity and imaginative endeavours forward. As we stand on the cusp of this transformative era, it's imperative we embrace automation not as a threat but as the ultimate collaborator, one that liberates us to reimagine the boundaries of what's possible in our work and beyond.

AI and Robotics in the DIY Factory As we delve into the transformative role of AI and robotics within the DIY factory model, it becomes increasingly clear that these technologies are not just supplemental; they're fundamental to the evolution of modern workspaces. Embracing AI and robotics opens a myriad of avenues for enhancing productivity, creativity, and personalization in projects, embodying the very essence of the DIY ethos while aligning perfectly with New Work principles.

The integration of AI in a DIY factory setting ventures beyond mere automation. It involves the utilization of intelligent systems capable of learning and adapting, thus fostering an environment where innovation flourishes. Imagine a workspace where AI algorithms predict project outcomes, suggest optimizations, and even collaborate with humans in real-time problem-solving. The potential for growth and learning in such an environment is boundless, echoing the continuous learning culture at the heart of New Work.

Similarly, robotics in the DIY factory transcends conventional manufacturing processes. Here, robotics offer not just precision and efficiency but also enable makers and creators to bring to life complex, intricately designed projects that would be impossible to realize manually. This synergy between human creativity and robotic precision embodies the DIY factory's mission to merge traditional craftsmanship with cutting-edge technology.

One of the key benefits of introducing AI and robotics into the DIY factory is the democratization of manufacturing. Small-scale creators gain access to tools and technologies once reserved for large-scale industrial environments. This shift not only levels the playing field but also encourages a culture of innovation and experimentation. Small businesses and individual entrepreneurs can prototype and produce with a speed and precision that accelerates the iterative design process, crucial for innovation.

The environmental implications of incorporating AI and robotics in DIY factories can't be overlooked. With precision comes reduced waste, and with efficient processes, there's less energy consumption. AI systems optimize resource use, ensuring that materials are utilized most effectively, which aligns with the sustainability goals central to many DIY factory endeavors and the broader New Work movement's emphasis on environmental responsibility.

However, the integration of these technologies also presents challenges, notably the steep learning curve and the need for ongoing education. Continuous learning becomes not just a benefit but a necessity. The DIY factory model, with its emphasis on community and shared knowledge, is uniquely positioned to address these challenges, championing a culture where learning and knowledge sharing are intrinsic values.

A critical aspect of successfully implementing AI and robotics is the fostering of a collaborative ethos. These technologies should be seen not as replacements for human workers but as partners. This collaborative approach is vital for innovation and is a core principle of both the DIY and New Work philosophies. By working alongside intelligent systems and robotics, individuals can focus on higher-level creative and problem-solving tasks, thereby enriching their work and personal growth.

The economic impact of embedding AI and robotics in DIY factories is profound. It enables small-scale operations to scale up efficiently, opening up new markets and creating opportunities for local economies to thrive. This technology-driven approach to scaling democratizes the manufacturing process, making it accessible and feasible for a broader spectrum of entrepreneurs.

Moreover, the role of AI and robotics in customizing products in the DIY factory setting is a game-changer. In a market increasingly driven by the desire for personalized products, these technologies allow for an unprecedented level of customization at scale. They enable creators to tailor products to individual needs and preferences without sacrificing efficiency or significantly increasing costs.

Implementing AI and robotics also necessitates a reevaluation of workplace safety and ethics. The DIY factory model, with its emphasis on community, offers a framework for developing best practices that ensure safety while fostering an ethical approach to technology use. It's about creating environments where technology serves humanity, enhancing both the creative process and the well-being of those involved.

Community involvement and feedback are pivotal in refining AI and robotics applications within the DIY factory. An open, collaborative environment invites input from a diverse range of users, ensuring that technology development is responsive to the needs and values of the community it serves. This participatory approach is in perfect harmony with New Work concepts, emphasizing autonomy, participation, and a collective pursuit of meaningful work.

Looking forward, the potential for AI and robotics in DIY factories to transform traditional industries is vast. By adopting these technologies, traditional businesses can revolutionize their operations, infusing innovation and creativity into processes that have remained unchanged for decades. This intersection of old and new symbolizes the essence of the DIY factory—a fusion of tradition with innovation, creating a future where work is both meaningful and sustainable.

The role of education in equipping individuals with the skills to navigate this new landscape cannot be understated. As AI and robotics become integral to the DIY factory, educational programs must evolve to include hands-on learning opportunities with these technologies. This approach not only prepares individuals for the future of work but also empowers them to be innovators and creators, embodying the spirit of both the DIY ethos and New Work.

In conclusion, the integration of AI and robotics into the DIY factory represents a convergence of technology, creativity, and human aspiration. It's a testament to the power of innovation to reshape not just how we work, but how we live, learn, and connect with each other. As we move forward, embracing these technologies within the DIY factory model offers a roadmap for building a future that values creativity, sustainability, and inclusivity—a future where work is not just about making a living, but about making a life worth living.

Technology as an Enabler of New Work Practices

In today's rapidly advancing world, technology stands at the forefront, not just as a tool, but as a powerful enabler of new, innovative work practices. Its pervasive influence is reshaping the very fabric of the workplace, blurring the lines between physical and digital, and between human and machine interactions. As we venture further into this dynamic landscape, it becomes imperative to understand how technology acts as a catalyst, fostering environments where creativity and efficiency coalesce to redefine what we consider 'work'.

At the heart of this transformative journey is the DIY factory model, a concept that embodies the ethos of innovation and autonomy. Here, technology is not just a facilitator but a partner in the creative process. From sophisticated AI algorithms that streamline operations to 3D printing technologies that bring ideas to life, these tools empower individuals and teams, allowing them to push the boundaries of what's possible.

Moreover, the integration of technology in work practices has democratized the means of production. No longer are sophisticated manufacturing techniques and data analytics the sole domain of large corporations. Small teams and even solo entrepreneurs now wield the power to analyze trends, prototype rapidly, and manufacture goods on a scale that was unimaginable just a few decades ago. This shift not only fuels innovation but also levels the playing field, enabling a more inclusive and diverse ecosystem of creators and innovators.

But the role of technology extends beyond merely amplifying productivity and creativity. It's also reshaping the very ethos of work environments, fostering cultures that value autonomy, flexibility, and continuous learning. Digital communication tools and project management platforms have made remote collaborations not just feasible but efficient, breaking down geographical barriers and enabling a global workforce. This unprecedented connectivity encourages cross-pollination of ideas, further invigorating the creative process.

Yet, embracing technology as an enabler requires a paradigm shift. Organizations and individuals alike must move away from viewing technology merely as a means to automate repetitive tasks. Instead, there is a need to see it as a foundational element in building a culture of innovation. This involves reimagining work processes, where technology and human ingenuity are intertwined, each amplifying the strengths of the other.

Adapting to this new era also means confronting the challenges head-on – from addressing the digital divide to ensuring ethical considerations are at the forefront of technological adoption. Fostering a culture of lifelong learning is essential, as the skills required in this new landscape evolve rapidly. Employers and educators must collaborate closely to equip workers with the skills to navigate and leverage these technological advancements.

In this context, leadership plays a crucial role. Leading in a technology-infused work environment demands not just a vision but a commitment to fostering an organizational culture that embraces change, values experimentation, and encourages risk-taking. True innovation leaders are those who empower their teams, giving them the tools and autonomy to experiment, fail, learn, and innovate.

One cannot overlook the role of environmental sustainability in the conversation about technology and new work practices. Here, technology serves not just as an enabler of efficiency and innovation but as a steward of our planet's resources. From optimizing energy consumption in manufacturing processes to leveraging data analytics for sustainable supply chains, technology offers pathways to a more sustainable and responsible approach to work and production.

The journey ahead is filled with opportunities and challenges in equal measure. Yet, history has shown us time and again that when faced with challenges, human ingenuity, powered by technology, finds a way forward. The future of work is not a distant horizon we're moving towards; it's a reality we're shaping today through every technological breakthrough and innovative work practice we embrace.

As we stand on the cusp of this new era, it's clear that technology's role transcends mere automation. It's an enabler of dreams, a bridge between imagination and reality, a tool that turns the impossible into the possible. The future of work is not just about working smarter or more efficiently; it's about working in ways we've yet to fully imagine, creating value that transcends markets and enriches societies.

Therefore, as we navigate this journey, let us leverage technology not just with a sense of purpose but with a vision. A vision that sees beyond the horizon, recognizing the untapped potential that lies in the synergistic relationship between humans and technology. This is not just the path to innovation and productivity; it's the road to a more creative, fulfilling, and sustainable future for all.

Chapter 8:
Building Blocks of Creativity and Innovation

In the compelling journey toward reshaping the landscape of modern work, a profound understanding of the core components that fuel creativity and innovation stands as a critical milestone. At the heart of disruptive change lie ideation and creative problem-solving, techniques that transcend conventional boundaries and foster an ecosystem of relentless exploration. It's within this fertile ground that the seeds of prototyping and experimentation find their nourishment, paving the way for groundbreaking advancements. Through a lens of inspiring resilience, we delve into the intricacies of transforming abstract ideas into tangible realities, a process empowering professionals and entrepreneurs alike to architect the future. This exploration isn't just about harnessing tools and techniques; it's a call to foster a mindset ripe for innovation, where barriers are viewed as mere stepping stones to the next breakthrough. As we navigate this chapter, we're not simply learning to create; we're embracing the very essence of innovation, setting the stage for a revolution in how we conceive, develop, and actualize visionary concepts. This narrative is more than a guide; it's a beacon for those daring to lead the charge toward a reimagined tomorrow, where creativity and innovation aren't just ideals but the very building blocks of a transformative new era of work.

Ideation and Creative Problem Solving

In the pulsating heart of innovation lies the critical process of ideation and creative problem solving, an essential pillar that supports the transformative architecture of creativity and innovation. True innovation is not just a spark but the result of rigorous brainstorming, relentless questioning, and imaginative exploration. It's about viewing challenges through a multidimensional lens, where problems are not roadblocks but stepping stones to groundbreaking solutions. This section delves into the strategic cultivation of ideation - that magical realm where questions like "What if?" and "Why not?" fuel the journey towards uncharted territories of creativity. Equipping oneself with a diverse toolkit of techniques and fostering an environment where every insight, no matter how small, is cherished, can transform mundane tasks into exciting opportunities for innovation. The ability to pivot from conventional thinking, blended with the courage to fail and learn from those failures, is what sets apart true innovators. In the DIY Factory model, where collaboration and adaptability reign, creative problem solving becomes not just an individual endeavor but a shared journey towards excellence. Embracing this dynamic process ensures that the path to innovation is one of constant evolution, filled with discoveries that propel both personal and communal growth.

Techniques and Tools As we navigate through the intricate web of innovation and creativity within the realm of DIY factories and New Work, it becomes imperative to focus on the techniques and tools that serve as the backbone of this transformative movement. In this segment, we'll embark on a journey to explore the plethora of methodologies and instruments that not only enhance productivity but also foster an environment ripe for creative exploration.

The essence of creativity within the New Work philosophy centers around a crucial component: ideation. Ideation, the process of generating a broad set of ideas without judgment, leverages tools such as brainstorming sessions, mind maps, and ideation workshops. These methods encourage divergent thinking, allowing individuals and teams to explore the width and breadth of possibilities before converging on actionable solutions. The physical or digital whiteboard becomes a canvas for imagination, where sticky notes or digital equivalents can represent the fleeting thoughts of innovators, ready to be organized, evaluated, and transformed into tangible concepts.

Beyond the generation of ideas, prototyping emerges as a critical step in the innovation process. Prototyping tools range from simple sketching and 3D modeling software to advanced CAD (Computer-Aided Design) systems and rapid prototyping machines like 3D printers. These tools allow innovators to quickly bring their ideas to life, testing form, function, and feasibility without significant investment in resources. This iterative process of creation, evaluation, and refinement is at the heart of the DIY factory ethos, encouraging a hands-on approach to problem-solving and design.

In tandem with physical prototyping tools, digital platforms play a crucial role in the collaborative aspect of innovation. Collaboration tools such as Slack, Trello, and Asana facilitate communication and project management within and across teams. These platforms enable a seamless flow of information, ensuring that every team member is aligned with the project goals and timelines. The use of such tools is reflective of the flat hierarchies characteristic of New Work, promoting openness, agility, and transparency in the pursuit of creative endeavors.

Emerging technologies such as AI (Artificial Intelligence) and robotics are also becoming instrumental in the DIY factory landscape. AI algorithms can assist in data analysis, pattern recognition, and even the ideation process, offering insights that may not be immediately apparent to the human mind. Robotics, on the other hand, extends the capabilities of human labor, automating repetitive tasks, and enabling more time for creative and strategic work. These technologies, when aligned with the human-centric approach of New Work, augment the creative capabilities of individuals and teams, leading to innovations that were once considered beyond reach.

Learning and experimentation are central to the continuous improvement and evolution of DIY factories. E-learning platforms, MOOCs (Massive Open Online Courses), and virtual reality (VR) training simulations serve as vital tools for skill development and knowledge sharing. They enable lifelong learning and adaptability among the workforce, ensuring that skills remain relevant in the rapidly evolving landscape of work and technology. By embracing these educational tools, organizations foster a culture of curiosity, resilience, and growth among their members.

The DIY factory model also leverages sophisticated inventory and supply chain management tools. Technologies such as IoT (Internet of Things) sensors and blockchain for traceability ensure a smooth operation by optimizing inventory levels, predicting maintenance needs, and ensuring the authenticity of materials. These tools not only contribute to efficient operations but also embed sustainability and ethical considerations into the fabric of the organization, aligning with the broader values of the New Work movement.

Visual communication tools such as Adobe Creative Suite and Canva democratize the ability to create compelling visual content, enabling anyone within the organization to express ideas visually, regardless of their graphic design skills. This empowerment enhances the quality of presentations and pitches, making the process of communicating ideas more effective and engaging.

Feedback and analytics tools, such as Google Analytics and Hotjar, provide insights into user behavior and preference, enabling a data-driven approach to innovation. By understanding the needs and behaviors of their end-users, organizations can refine products and services to better meet market demands, ensuring that their innovations are not only creative but also relevant and impactful.

To conclude, the blend of techniques and tools available to those navigating the New Work landscape offers a powerful toolkit for transforming ideas into reality. From fostering the initial spark of creativity through ideation to bringing concepts to life via prototyping, and from enhancing collaboration to facilitating continuous learning, these tools and techniques are pillars upon which the future of work rests. Embracing these resources, individuals and organizations alike empower themselves to navigate the complexities of the modern world, driving innovation, fostering creativity, and ultimately contributing to a more vibrant, dynamic, and inclusive future for all.

Prototyping and Experimentation

In the journey of transforming innovative ideas into tangible realities, prototyping and experimentation stand as indispensable cornerstones. These processes not only breathe life into abstract concepts but also serve as crucial feedback loops, enabling creators to iterate and refine their creations with precision and insight. In the realm of modern work culture, where the DIY factory and New Work principles collide, prototyping and experimentation are not just methodologies but a mindset that fuels continuous improvement and innovation.

Imagine stepping into a workspace where the air buzzes with creativity, where every corner is teeming with ideas waiting to take form. This is where prototyping comes into play. Prototyping is the art of creating preliminary models of products or concepts. It's about making ideas tangible, accessible, and most importantly, testable. It breaks down monumental visions into manageable, testable parts, making the daunting task of bringing a novel idea to life suddenly achievable.

Experimentation walks hand in hand with prototyping. It's the systematic process of trying out different prototypes, pushing boundaries, and challenging preconceived notions about what works and what doesn't. Through experimentation, we invite serendipity and innovation, discovering solutions to problems we didn't even know existed. It's a testament to the ethos that failure isn't a setback but a pivotal step towards uncovering groundbreaking ideas.

Consider the story of a successful startup that began its journey in a modest garage, armed with nothing but a vision and an unyielding drive to innovate. The founders, embodying the DIY spirit, embarked on relentless cycles of prototyping and experimentation. Each version of their product, no matter how rudimentary, provided invaluable insights that informed the next iteration. This iterative process not only honed their product to meet the market's needs but also instilled a culture of agile development and resilience within their team.

In a professional environment that values innovation, the ability to rapidly prototype and experiment provides a competitive edge. It allows teams to navigate the uncertain waters of innovation with confidence, making informed decisions based on tangible data rather than mere speculation. The feedback gathered from each prototype iteration is a goldmine of insights, driving product development towards success with greater precision.

Yet, the journey of prototyping and experimentation is not devoid of challenges. It requires a shift in mindset, from fearing failure to embracing it as an essential step towards innovation. It necessitates an environment that supports risk-taking, where the pursuit of innovation is championed over the comfort of sticking with the status quo. Creating such an environment is pivotal for fostering a culture that thrives on experimentation.

Forward-thinking companies have realized the immense potential of prototyping and experimentation. They invest in state-of-the-art laboratories, maker spaces, and innovation hubs where their teams can freely explore and innovate. These spaces are not just physical locations but symbolic of a company's commitment to fostering creativity and innovation at its core.

As we delve deeper into the digital age, the tools and technologies available for prototyping and experimentation have evolved dramatically. From advanced 3D printing technologies to virtual reality simulations, the arsenal of tools at our disposal has expanded, enabling more complex and sophisticated prototypes to be developed faster and more efficiently than ever before.

However, the essence of prototyping and experimentation transcends the physical realm. It embodies a philosophy of iterative learning, a continuous cycle of testing, learning, and improving. It's a framework that can be applied not only in product development but in solving complex problems across various domains. Whether it's refining a business process or creating a new service model, the principles of prototyping and experimentation can lead the way.

In implementing these practices, it's crucial to cultivate a diverse and inclusive team. Diversity in thought, background, and expertise enriches the experimentation process, bringing in a multitude of perspectives that can lead to truly innovative solutions. It's the blend of different viewpoints and experiences that ignites creativity, making the process of innovation a collective and inclusive journey.

The magic of prototyping and experimentation lies in their ability to make the impossible seem possible. They are the building blocks of creativity and innovation, turning ideas into reality, one iteration at a time. By embracing these practices, professionals, entrepreneurs, and enthusiasts alike can embark on a fulfilling journey of discovery and innovation.

In conclusion, the path towards innovation is paved with trial and error, learning and evolving. Prototyping and experimentation are not mere steps in the product development process but a mindset that celebrates curiosity, resilience, and the relentless pursuit of excellence. They are the heart and soul of a culture that seeks to break the mold and chart new territories in the ever-evolving landscape of work and creativity.

So, let us embrace these practices with open arms and an open mind. Let's prototype, experiment, and innovate, for it's through these endeavors that we can unlock our full potential and bring forth innovations that not only serve our immediate needs but also contribute to a brighter, more creative future for all.

In the grand tapestry of modern work culture, where the threads of creativity, innovation, and the DIY ethos intertwine, prototyping and experimentation emerge as the vibrant colors that bring the masterpiece to life. They remind us that the journey of innovation is not a linear path but a vibrant mosaic of experiences, learning, and growth. And it is within this vibrant mosaic that we find the true essence of creativity and innovation.

Chapter 9:
The Economic Impact of DIY
Factories and New Work

In the age where innovation intersects with the allure of the do-it-yourself (DIY) spirit, we embark on a journey to unravel the profound economic impact of DIY factories coupled with the principles of New Work. As we shift from mere consumerism to active production, it becomes clear that the traditional economic paradigms are being challenged and reshaped. This transformative movement towards engaging, self-sufficient production methods not only democratizes the means of production but also fosters a dynamic environment for creativity and innovation to flourish. The infusion of New Work principles into this mix further revolutionizes our approach to work, emphasizing autonomy, purpose, and collaboration over hierarchical, rigid structures. Through this lens, we witness a burgeoning economic model that supports the growth of DIY factories, enabling them to thrive as incubators of innovation and creativity. Moreover, the global reach of these principles signals a promising horizon where new work methodologies are not just localized experiments but part of a global shift towards more meaningful, fulfilling, and sustainable practices. As forward-thinking professionals, entrepreneurs, and enthusiasts dive deeper into the essence of DIY factories and New Work, it becomes imperative to embrace these paradigms, fostering an economic landscape that is not only prosperous but also equitable and inclusive.

Shifting from Consumerism to Production

In a world increasingly characterized by consumption, the advent of DIY factories and the principles of New Work are heralding a seismic shift towards production, empowering individuals and communities to take control of their economic destinies. This movement is not merely about the physical creation of goods but represents a profound transformation in how we perceive value, engage with our work, and contribute to the economy. By embracing the tenets of DIY culture, we not only forge paths toward self-sufficiency but also stimulate local economies, fostering innovation and job creation in the process. The shift from consumerism to production challenges traditional economic models, paving the way for a more inclusive and resilient future. As we realign our role in the economy from passive consumers to active producers, we unlock the potential for a more equitable and sustainable world, demonstrating that with creativity, collaboration, and a bit of ingenuity, we can redefine what it means to work and to thrive in the 21st century.

Economic Models Supporting DIY Factories

In the preceding sections, we uncovered the transformative synergy between DIY factories and New Work practices. However, the true catalyst behind their widespread adoption and success lies in the innovative economic models that support them. These models not only enhance sustainability but also foster community engagement, creativity, and growth.

At the core of DIY factories is the principle of decentralization. This approach challenges traditional manufacturing and work paradigms by distributing production processes across many small-scale, agile entities rather than concentrating them within large, monolithic corporations. Economically, this has profound implications. It democratizes the means of production, making it accessible to a broader spectrum of individuals and communities. This accessibility is not just about physical resources but also encompasses knowledge, skills, and networks, thereby fueling a more inclusive economy.

The economic models that bolster DIY factories often involve shared resources and collaborative consumption. Coworking spaces, makerspaces, and tool libraries exemplify how resources can be pooled to reduce individual costs, foster a sense of community, and minimize environmental impact. This sharing economy aspect is pivotal, as it ensures that high-quality tools and facilities are within reach for more aspiring creators and entrepreneurs.

Additionally, many DIY factories thrive through a hybrid of revenue models that blend traditional sales, subscription services, and community funding such as crowdfunding. This multifaceted approach not only provides financial stability but also engages the community directly in the success of the enterprise. It creates a feedback loop where products and services can be rapidly iterated based on real user feedback, leading to a more user-driven innovation process.

The rise of digital platforms has significantly aided the economic viability of DIY factories. Online marketplaces, social media, and crowdfunding platforms offer powerful tools for marketing, sales, and community building. They allow DIY factories to reach a global audience, test ideas before committing significant resources, and raise funds for expansion. This global connectivity also facilitates the sharing of ideas and best practices, accelerating innovation and growth.

Another cornerstone of the economic model supporting DIY factories is the emphasis on sustainability and circular economy principles. By prioritizing the use of recyclable materials, promoting repair and reuse over disposal, and focusing on sustainable sourcing, DIY factories not only reduce their environmental impact but also often realize cost savings and build brand loyalty among environmentally conscious consumers.

Moreover, the educational component of DIY factories presents another economic avenue. Workshops, courses, and mentorship programs not only serve as an additional revenue stream but also help to cultivate a skilled workforce tuned to the principles of DIY and New Work. This investment in human capital not only benefits the individual but also contributes to a more dynamic and resilient local economy.

The gig economy and flexible work arrangements also play a crucial role in the economic models of DIY factories. By employing freelancers and part-time workers, DIY factories can maintain agility and scale labor according to demand. This flexibility is beneficial for both the factory, which can operate more efficiently, and for workers, who seek autonomy and balance between their professional and personal lives.

Public-private partnerships can further enhance the economic model of DIY factories. Government support through grants, tax incentives, and regulatory flexibility can significantly lower the barriers to entry for new makerspaces and DIY factories, sparking innovation and economic growth at the local level.

Importantly, the success of DIY factories often hinges on the strength of the community they cultivate. A strong, engaged community not only provides a customer base but also fosters a network of mentors, collaborators, and brand ambassadors. This social capital is invaluable, as it drives word-of-mouth marketing, attracts talent, and builds a supportive ecosystem that can weather economic downturns.

However, the economic sustainability of DIY factories is not without challenges. Issues such as access to initial capital, managing the complexity of shared resources, and navigating the regulatory landscape can be significant hurdles. Yet, the continued evolution of collaborative finance options, digital technologies, and supportive policy frameworks are paving the way for more resilient economic models.

The future is bright for DIY factories, as they are positioned at the intersection of innovation, sustainability, and community. Their economic models, which emphasize accessibility, collaboration, and flexibility, are not merely pathways to financial viability but are catalysts for a more inclusive, creative, and sustainable future of work and production.

In conclusion, the economic models supporting DIY factories represent a radical shift from conventional business practices. They challenge us to reimagine the possibilities of how we produce, work, and engage with our communities. As we move forward, embracing these models will not only fuel the growth of DIY factories but also contribute to building a more adaptive, resilient, and human-centric economy.

As we continue our journey through this book, keep these economic insights in mind. They are not just theoretical concepts but practical blueprints for anyone looking to implement or participate in the DIY factory movement. The shift toward more decentralized, collaborative, and sustainable economic models is not just possible; it's already underway, and by engaging with these ideas, we can all be part of shaping this exciting future.

The Global Reach of New Work Principles

In the tapestry of modern economies, the principles of New Work serve as vibrant threads, weaving through the fabric of societies around the globe. This global reach is not just a testament to the adaptability and appeal of New Work ideologies but also signals a profound shift in how work is conceptualized, organized, and executed across different cultures and economies. As we delve into the worldwide impact of these principles, it's crucial to recognize the diverse ways in which communities, companies, and countries are engaging with and transforming these ideas to fit their unique contexts.

The essence of New Work principles lies in their emphasis on flexibility, autonomy, creativity, and the pursuit of meaningful and fulfilling work. These values resonate with a growing number of people who seek not just to make a living but to make a life through their work. The digital revolution, with its decentralizing force, has democratized access to tools and platforms that enable these principles to be lived and implemented on a scale never before possible.

One remarkable aspect of New Work's global spread is its adaptability. In Scandinavian countries, for example, New Work principles align seamlessly with societal values of equality, balance, and collective well-being. Here, governments and organizations have been pioneers in implementing flexible working hours, comprehensive remote work policies, and flat organizational structures that empower individuals and promote work-life harmony.

Meanwhile, in the bustling tech hubs of Asia, the principles of New Work are fueling a culture of innovation and entrepreneurship. In cities like Bangalore, Seoul, and Beijing, dynamic co-working spaces and DIY factories serve as incubators for startups and freelancers who embody the New Work ethos. These environments encourage experimentation, rapid prototyping, and a fail-fast mentality that is vital for cutting-edge innovation.

In the emerging economies of Africa and Latin America, New Work principles are being leveraged to address unique challenges and opportunities. Through mobile technology and the internet, remote work has become a gateway to the global economy for many, enabling access to education, markets, and networks that were previously out of reach. Here, New Work is not just about personal fulfillment but about economic empowerment and community development.

Further, the global reach of New Work has been catalyzed by the rise of the gig economy and freelance platforms that connect talent with opportunities, regardless of geographic boundaries. This has created a fluid and dynamic global marketplace where skills, ideas, and creativity are the currencies of exchange. The global network of creatives, programmers, designers, and other professionals embodying New Work principles is a testament to the movement's inclusive and boundary-crossing nature.

However, the global embrace of New Work also presents challenges, particularly in terms of ensuring equitable access and mitigating the risks of exploitation and instability associated with gig and remote work. Addressing these challenges requires collaborative efforts between governments, organizations, and the New Work community to create supportive policies, infrastructures, and cultures that uphold the values of fairness, security, and inclusivity.

The intersection of New Work with social and environmental sustainability worldwide is another critical dimension. By encouraging practices such as telecommuting, digital collaboration, and the use of sustainable materials in DIY factories, New Work principles contribute to reducing carbon footprints and promoting responsible consumption and production. This synergy between New Work and sustainability goals showcases the movement's potential to address some of the most pressing global challenges of our time.

Education systems around the world are also beginning to reflect and incorporate New Work principles, recognizing the need to prepare students for a future where creativity, critical thinking, and adaptability are paramount. From project-based learning to initiatives that blur the lines between classrooms and workplaces, educational institutions are evolving to foster the skills and mindsets aligned with New Work values.

The movement's global reach is further amplified by digital platforms and social media, which facilitate the sharing of ideas, resources, and best practices across borders. This global community of New Work advocates serves as a powerful force for innovation and change, inspiring others to reimagine work and create environments where everyone has the opportunity to pursue their passions and potential.

As we consider the future, it's clear that the principles of New Work will continue to spread and evolve, shaped by local cultures, economic conditions, and technological advancements. This dynamic and iterative process is not without its challenges, but it holds the promise of a more inclusive, creative, and sustainable global work culture.

In conclusion, the global reach of New Work principles is a testament to the universal desire for work that is not only productive but also meaningful, flexible, and fulfilling. By embracing these values, individuals, organizations, and societies worldwide are crafting a new narrative for work—one that prioritizes personal well-being, community engagement, and sustainable growth. As this movement grows, it will undoubtedly continue to transform and be transformed by the diverse communities that embrace it, paving the way for a future where work enriches not just economies, but lives and the planet as well.

The journey of integrating New Work principles into our global fabric is ongoing, promising a future where work is not just a means to an end but a key contributor to a fulfilling and sustainable life. For forward-thinking professionals, entrepreneurs, and enthusiasts, the global reach of New Work offers a wide canvas of opportunities to innovate, influence, and inspire. Embracing these principles is not merely a professional choice but a commitment to being part of a global movement reshaping the future of work for generations to come.

Chapter 10:
Legal and Ethical Considerations

As we steer into the realms of DIY factories and New Work, the conversations invariably pivot toward the uncharted territories of legal and ethical considerations. Anchored in the idea that innovation must coexist with integrity, this chapter delves deep into the intricacies of intellectual property amidst a sharing economy, spotlighting the delicate balance between collaboration and ownership. The surge of open-source platforms has democratized innovation, yet it begs the question - how do we navigate these shared spaces respectfully and legally? Furthermore, the gig economy, with its flexibility and autonomy, presents its own ethical mazes to traverse. In ensuring fair practices, we must be vigilant about not just the legality but the humanity of our operations. Reflect on this: as forward-thinkers and trailblazers, it's incumbent upon us to foster environments where ethical considerations are not just an afterthought but a cornerstone of our innovation ecosystem. By laying a strong ethical foundation, we not just safeguard our ventures but also champion a new era where progress and integrity stride hand in hand towards a horizon of limitless possibilities.

Intellectual Property in a Sharing Economy

In the vibrant landscape of a sharing economy, the paradigms of intellectual property (IP) undergo a transformative shift, facilitating an environment where innovation thrives on collaboration rather than competition. As we navigate through this evolving terrain, it becomes imperative to understand that traditional notions of IP ownership can sometimes stifle creativity and impede the collective advancement of technology and ideas. By fostering a culture that celebrates open source projects and collaborative efforts, we not only accelerate the pace of innovation but also ensure that the fruits of this progress are accessible to a wider community. This approach not only aligns with the ethos of the DIY factory and New Work movement but also challenges us to rethink how we value and protect creative work. The sharing economy prompts us to consider a balance where protecting inventors' rights and encouraging a free exchange of ideas coexist, thereby driving a future where innovation is unbounded and inclusively shared across borders and industries.

Open Source and Collaboration In the unfolding landscape of modern work culture, an undercurrent that's shaping the very fabric of innovation and creativity is the open source movement. Engaging in open source projects and fostering collaboration are not just about sharing code or ideas freely but about tapping into a collective intelligence that propels breakthroughs and accelerates progress. This ethos, deeply embedded in the DIY factory and New Work concepts, serves as a foundation for a culture where sharing, co-creation, and mutual progress become the norm rather than the exception.

Open source initiatives challenge the traditional gatekeepers of knowledge and innovation by democratizing access. This movement has proven that when people come together, unshackled by the barriers of ownership and proprietary constraints, they can create solutions that are not only innovative but often outperform closed, competitive efforts. In this realm, collaboration across borders, disciplines, and industries leads to a cross-pollination of ideas that fosters unanticipated innovations.

The ethos of open source aligns perfectly with the principles of New Work, which advocates for autonomy, freedom, and meaningful engagement in professional endeavors. Integrating open source practices within organizations encourages a culture of transparency and shared goals. It inspires individuals to contribute their best work, not just for personal gain or recognition, but for the advancement of the entire community or project at hand. This environment is conducive to nurturing a sense of ownership and responsibility in each contributor, driving elevated levels of motivation and commitment.

Moreover, the collaborative nature of open source projects facilitates a de facto form of continuous learning and skill development. As individuals engage with a global community of professionals, hobbyists, and enthusiasts, they are exposed to diverse perspectives, techniques, and challenges. This exposure not only hones their existing skills but also compels them to acquire new ones, making the process a powerful engine for professional growth and innovation.

Implementing open source and collaboration within the framework of a DIY factory can significantly amplify its impact. By leveraging collective intelligence, DIY factories can solve complex problems more efficiently, innovate at a faster pace, and reduce development costs. This inclusive approach also enables these hubs of creativity to remain at the cutting edge of technology and trends, as the open exchange of information keeps them aligned with the latest advancements and community insights.

One of the most compelling arguments for embracing open source is its role in fostering inclusivity and diversity. By lowering the barriers to entry, it invites participation from individuals of varied backgrounds, experiences, and skills. This diversity enriches the ecosystem, introducing a wide array of perspectives that can challenge conventional thinking and spur innovative solutions. The emphasis on collaboration and collective achievement helps to create a supportive environment where all contributors feel valued and empowered.

Yet, navigating the world of open source and collaboration is not without challenges. Issues related to intellectual property, quality control, and sustainability of efforts arise. Overcoming these requires a thoughtful approach that includes establishing clear guidelines, encouraging respectful and constructive interactions, and devising models that ensure the longevity and health of projects. This balancing act between openness and structure is crucial for harnessing the full potential of collaborative innovation.

The integration of open source principles also necessitates a shift in mindset at both the organizational and individual levels. Shifting from a competition-driven to a collaboration-centric model challenges deeply ingrained beliefs about success and ownership. Cultivating a culture that values open contributions and collective outcomes over individual accolades requires patience, leadership, and a steadfast commitment to the core values of openness and community.

Success stories of open source and collaborative projects abound, serving as testament to their potential. From software like Linux and Apache to collective knowledge endeavors like Wikipedia, these examples highlight the remarkable achievements possible when communities unite around shared objectives. These successes offer invaluable lessons and inspiration for those seeking to incorporate these principles into their DIY factories and innovative workspaces.

Moreover, the open source movement has spurred the development of tools and platforms that facilitate collaboration. Platforms such as GitHub, GitLab, and others not only provide the infrastructure for sharing projects and code but also foster communities around them. These tools make contributing to projects more accessible and managing collaboration more straightforward, further fueling the open source and collaborative spirit.

The act of sharing and collaboration extends beyond just projects and code; it encompasses the sharing of experiences, challenges, and best practices. Forums, conferences, and workshops dedicated to open source principles offer rich opportunities for networking, learning, and inspiration. Engaging in these communities not only aids in professional development but also strengthens the bonds within the open source ecosystem, creating a supportive network that can be leveraged for guidance and collaboration.

The principles of open source and collaboration are increasingly being recognized as critical drivers of innovation in the commercial sphere as well. Companies, big and small, are beginning to understand the value of engaging with open source communities and adopting more collaborative and transparent practices internally. This shift not only enhances their innovation capabilities but also aligns them more closely with the values and expectations of a new generation of workers and consumers.

For initiatives rooted in the DIY factory and New Work concepts, embracing open source and collaboration offers a path to not just innovation, but also resilience, adaptability, and community engagement. These principles provide a framework for sustainable growth, enabling initiatives to thrive in an ever-changing landscape. They remind us that in the new era of work, success is not just about what we create, but how we create it—and with whom we share the journey.

As we forge ahead, the fusion of open source and collaboration stands as a beacon for those aspiring to shape a future that values creativity, inclusivity, and shared progress. It's an invitation to join a global movement that transcends traditional boundaries and limitations, offering a glimpse into a world where collective efforts lead to unprecedented levels of innovation and impact. By embracing these principles, we not only contribute to the evolution of work and creativity but also participate in building a more open, connected, and equitable world.

Navigating the Gig Economy Ethically

In our journey through the evolution and future of work, a significant pitstop is the ethical navigation of the gig economy. This sector, characterized by short-term contracts or freelance work as opposed to permanent jobs, has burgeoned with the rise of digital platforms. Yet, as it grows, so does the complexity of its ethical landscape.

The gig economy epitomizes flexibility and autonomy, offering individuals the chance to work on their terms. However, this freedom often comes without the safety nets traditionally provided by employment. Therein lies the first ethical challenge: ensuring that gig workers are not exploited but rather empowered by this new work paradigm.

Empowerment in the gig economy means enabling workers with fair compensation, reasonable working hours, and access to benefits. It's about dismantling the notion that flexibility equates to instability. Achieving this requires concerted efforts from all stakeholders— platforms, workers, regulators, and consumers.

Digital platforms that facilitate gig work have a pivotal role in shaping its ethical foundation. They can pioneer innovative ways to offer benefits traditionally reserved for employees, like health insurance or retirement plans, setting a new standard for gig worker support. Transparent algorithms for job assignments and fair dispute resolution mechanisms are further steps towards an ethical platform model.

Regulation is another critical piece of the puzzle. Laws and policies need to evolve alongside work paradigms to protect workers' rights without stifling the innovation that the gig economy brings. This includes rethinking labor laws to recognize the unique nature of gig work and ensuring that gig workers have a voice in these discussions.

From the perspective of gig workers themselves, ethical navigation involves understanding their rights and advocating for themselves. It means engaging in continuous learning and skill development to enhance their value and employability in a competitive market. It's about forming communities and networks to share information and support one another.

Consumers, too, play a role in the ethical gig economy. By choosing platforms and services that treat workers fairly, consumers can drive demand for ethical practices. Their choices can support businesses that are committed to responsible gig work models.

Moreover, transparency is vital for all actors in the gig economy. Platforms must be transparent about how they calculate pay and assign work, just as workers must be transparent about their qualifications and availability. Regulators must also be transparent about the rules and their implementation. This transparency fosters trust among all parties and facilitates ethical decision-making.

Another aspect to consider is the international dimension of the gig economy. With digital platforms offering global access to work opportunities, ethical considerations extend across borders. This includes respect for cultural differences and ensuring that global work opportunities do not lead to a 'race to the bottom' in terms of labor standards.

In navigating the gig economy ethically, we must also consider the environmental impact. The digital nature of many gig jobs can reduce the need for commuting, potentially lessening environmental footprints. However, this advantage must be weighed against the environmental costs of digital infrastructure and the gig economy's contribution to the consumption-driven model.

Innovation in governance and business models can further support ethical practices. Social enterprises and cooperatives, for example, can offer models for gig work that prioritize worker well-being and community benefits over profit.

Educational initiatives are crucial in preparing both workers and employers to thrive ethically in the gig economy. From understanding new labor laws to embracing sustainable business practices, education can empower all participants to make informed and ethical decisions.

Finally, the ethical navigation of the gig economy demands a future-focused mindset. It's about anticipating the changes and challenges that lie ahead and proactively addressing them. This involves not just adapting to the gig economy, but actively shaping it to be more inclusive, fair, and sustainable.

In conclusion, navigating the gig economy ethically is a multifaceted endeavor. It requires the collaboration of all stakeholders to forge a path that balances innovation with protection, autonomy with security, and flexibility with fairness. As we continue to shape the future of work, let us commit to principles that uplift and support every participant in the gig economy.

Embracing the gig economy ethically is not merely a challenge; it's an opportunity to redefine work in a way that enriches humanity. By fostering an ecosystem that values ethical practices, we can ensure that the gig economy becomes a force for good, offering a fulfilling and equitable work experience for all.

Chapter 11:
Education and Skill Development for the Future

In an epoch where innovation and agility stand as the pillars of professional success, "Education and Skill Development for the Future" delineates a visionary path for crafting the workforce of tomorrow. Emphasizing a paradigm shift from traditional rote learning to a dynamic, hands-on approach, this chapter heralds the advent of a learning ecosystem where creativity and innovation are not just encouraged but are indispensable. Imagine transforming the classical classroom into an imaginative workshop, where ideas flourish, skills are honed in real-time, and the lines between learning and doing blur. This isn't just about redefining the spaces we learn in; it's about fostering a culture of lifelong learning and skill sharing that evolves with us, ensuring we're not just prepared but ahead of the curve in an ever-changing professional landscape. By rethinking educational frameworks and embedding the principles of DIY factories and new work methodologies into our learning institutions, we imbue future generations with the resilience, creativity, and adaptability needed to thrive. Whether you're an educator shaping the minds of tomorrow, a professional navigating the complexities of a shifting job market, or an entrepreneur looking to foster a culture of continuous innovation, this chapter is your blueprint for building a future where the pursuit of knowledge and the development of skills are ceaselessly intertwined with the DNA of success.

Rethinking Education for Creativity and Innovation

In an era where traditional education models are being outpaced by the rapid evolution of technology and societal needs, it's imperative to initiate a fundamental shift in how we perceive and impart education. This pivotal transformation revolves around redefining the core objectives of learning environments to nurture creativity and innovation. As we delve deeper into the significance of fostering a culture of dynamic learning and skill development, it becomes clear that education shouldn't just be about the dissemination of knowledge but about inspiring a relentless pursuit of exploration and experimentation. By integrating hands-on workshops, collaborative projects, and real-world problem-solving into the curriculum, educators can unlock the immense potential of students, preparing them to not only thrive in but also shape the future of work. This approach empowers learners to view challenges as opportunities for growth, encouraging a mindset that values continuous learning and adaptability. Equipping individuals with the tools to navigate the complexities of the modern workforce is not just an educational imperative but a societal one, ensuring that we are cultivating a generation capable of leading the charge in innovation, sustainability, and global progress.

From Classroom to Workshop The journey from traditional education formats to dynamic, hands-on workshop environments signals not just a cultural shift, but a reinvention of learning itself. In the realms of innovation and creativity, the transition is not merely physical—from a room of desks to a space filled with tools—it's profoundly philosophical. It demands we reevaluate not just what we learn, but how we absorb, apply, and iterate on knowledge.

In this transformative landscape, the classroom is no longer the sole custodian of education. The workshop becomes an arena of collaborative exploration, where mistakes are not just permitted;

they're expected, analyzed, and valued. This shift isn't just beneficial; it's imperative. In a world that's rapidly changing, the ability to adapt, experiment, and innovate is not just an advantage. It's survival.

Consider the workshop not just as a place, but as an ecosystem. It's where ideas meet materials, tools, and technologies, bridging thoughts with tangible outcomes. This hands-on approach encourages a deeper understanding of concepts, as learners move from abstract theories to concrete applications. It fosters a culture of "doing" rather than merely "knowing," which is critical in cultivating the skills required for the future's uncertain challenges.

Embedded within the workshop model is the principle of collaborative learning. Unlike traditional classroom settings, where the flow of knowledge is predominantly unidirectional, from teacher to student, workshops thrive on multi-directional learning flows. Everyone, regardless of expertise, learns from each other. This peer-to-peer exchange democratizes knowledge and empowers participants, fostering a community where everyone contributes, challenges, and grows together.

The transition from classroom to workshop also places a premium on creativity and innovation. In a workshop, the focus shifts from memorizing answers to posing questions. "What if?" becomes more crucial than "What is?" This environment encourages learners to venture beyond the confines of conventional solutions, exploring new possibilities and prototyping their ideas without the fear of failure. After all, every failure is a lesson leading to innovation.

The integration of technology within workshops accelerates this learning transformation. From 3D printers to programmable microcontrollers, technology not only expands what can be created in these spaces but also reshapes the learning experience. Learners can design, test, and tweak prototypes rapidly, facilitating a faster, iterative learning process. This immediacy bridges the gap between idea conception and realization, making innovation not just accessible but also immediate.

Yet, acknowledging the power of workshops doesn't diminish the value of traditional classrooms. Instead, it invites a fusion where theoretical frameworks meet experimental hands-on learning. The most potent educational experiences emerge when learners can oscillate between understanding the theoretical underpinnings and applying them practically. This hybrid model ensures a holistic understanding, preparing learners for the nuances of real-world problem-solving.

Moreover, the workshop model champions the ethos of lifelong learning. It acknowledges that in an ever-evolving world, the acquisition of knowledge cannot be confined to the early decades of life. By making learning continuous, contextual, and community-driven, workshops exemplify how education can adapt to the needs of an age where change is constant.

This evolution also requires a reimagining of the educator's role. From authoritative knowledge holders, educators become facilitators and co-learners in the workshop environment. They guide, inspire, and challenge, creating a learning experience that's tailored to the individual's curiosity and the group's dynamic needs. This shift not only enriches the learning experience but also rekindles educators' passion for teaching.

In transitioning from classroom to workshop, it's also crucial to recognize the diversity of learners. Workshops, with their emphasis on practical skills and collaborative learning, can be particularly inclusive for individuals who might not thrive in traditional academic settings. By catering to different learning styles and empowering learners to take charge of their education, workshops can level the educational playing field, offering multiple paths to success.

The measure of success in workshops also diverges from traditional norms. Success isn't gauged by the ability to regurgitate information during exams but by the capacity to apply knowledge to solve complex problems, to learn from failure, and to persistently innovate. This redefinition of success aligns more closely with the demands of the contemporary world, where adaptability, creativity, and resilience are paramount.

Implementing the shift from classroom to workshop isn't without its challenges. It requires a reevaluation of educational standards, curricula, and assessment methods. It demands investment in resources, training for educators, and a cultural shift towards valuing hands-on, experiential learning. Yet, the benefits—a more engaged, capable, and innovative generation—are undeniable.

For professionals, entrepreneurs, and enthusiasts at the intersection of modern work culture and innovation, embracing the workshop model is more than an educational choice; it's a strategic imperative. The skills honed in workshops—collaboration, creativity, and technical proficiency—are the same skills that drive the DIY factory model and underpin new work paradigms.

In the end, the journey from classroom to workshop is more than a pedagogical shift. It is a reflection of a broader transformation in how society understands learning, work, and innovation. By fostering environments where learners are empowered to explore, create, and innovate, we're not just preparing them for the future; we're helping to shape it.

The fusion of education and hands-on experience epitomized by the workshop model isn't just a response to the changing landscape of work and technology. It's a proactive step towards a future where learning and doing are inseparable, fostering a generation not just of thinkers but of makers, doers, and innovators. This is the essential evolution required to navigate and thrive in the complexities of the

21st century. And the transition from classroom to workshop is just the beginning.

Lifelong Learning and Skill Sharing

In a world that's constantly evolving, the value of lifelong learning and skill sharing cannot be overstressed. This concept isn't merely a trend but a foundational element in fostering environments where creativity and innovation flourish. At the heart of the DIY factory ethos and new work paradigms is an unyielding belief in the power of continuous education and the sharing of knowledge. Why? Because today's society demands adaptability, and the ability to learn, unlearn, and relearn is indispensable.

Consider the pace at which technology evolves. What was groundbreaking yesterday may become obsolete tomorrow. Professionals, entrepreneurs, and enthusiasts within the DIY and New Work movements recognize this. They embrace a culture of learning not as a chore but as a continuous, exciting journey. This mindset shift transforms traditional notions of education from a finite phase of life into an ongoing process that enriches both personal and professional dimensions.

But what does lifelong learning look like in practice? It diverges significantly from conventional education systems. Rather than structured, curriculum-based learning, it leans towards experiential, hands-on learning experiences. This shift from the classroom to the workshop, as mentioned in the previous chapter, underscores a more pragmatic approach to skill acquisition.

Lifelong learning in this context also emphasizes versatility. It's about broadening one's skill set not just vertically in one's area of specialization, but also horizontally across a variety of disciplines. This polymath approach prepares individuals to think more holistically and innovatively, enabling them to draw connections between seemingly unrelated fields.

Equally significant is the concept of skill sharing, which naturally complements lifelong learning. The new work culture, especially within DIY factories, thrives on collaboration. Here, knowledge is not hoarded but shared freely. This ethos fosters a community where individuals learn from each other, pooling their expertise to push creative boundaries and solve problems more efficiently.

Technology plays a pivotal role in facilitating this culture of learning and sharing. Online platforms and social media have democratized access to knowledge, allowing anyone with an internet connection to learn new skills or share their expertise with a global audience. This has led to an explosion of creativity and innovation, as ideas and techniques are freely exchanged across borders.

However, fostering an environment that prioritizes lifelong learning and skill sharing requires more than just access to technological tools. It demands a cultural shift within organizations and communities. This shift involves reimagining leadership roles, flattening hierarchies to encourage open communication, and implementing systems that reward collaboration and knowledge sharing.

The benefits of embracing lifelong learning and skill sharing are multifaceted. For individuals, it enhances employability, adaptability, and personal satisfaction. For organizations, it leads to a more skilled, innovative, and resilient workforce. And on a societal level, it drives economic growth and advances human progress.

Yet, challenges remain. The most significant of these is overcoming entrenched educational and workplace cultures that prioritize competition over collaboration and standardization over creativity. It requires concerted effort from all stakeholders - educators, employers, policymakers, and learners themselves - to effect this cultural transformation.

Success stories from within the DIY factory and new work movements provide valuable lessons. These narratives highlight how environments that prioritize continuous learning, open sharing of knowledge, and collaboration can lead to remarkable innovations and breakthroughs. They serve as inspiration and a roadmap for others looking to cultivate similar ecosystems.

For individuals looking to thrive in the future of work, the message is clear: Embrace lifelong learning, share your skills, and remain curious and adaptable. This approach is not only beneficial but essential in navigating the complexities of modern work landscapes.

Organizations, too, must adapt. By fostering cultures that prioritize ongoing education and skill sharing, they can unlock the creative potential of their teams and navigate the challenges of tomorrow more successfully.

In conclusion, the paradigms of lifelong learning and skill sharing are central to the DIY factory ethos and new work movements. They reflect a profound understanding that in a world where change is the only constant, our ability to learn continuously and share our knowledge freely is not just an advantage but a necessity. By adopting these principles, we prepare ourselves and our communities for a future that's not only more innovative and productive but also more fulfilling and resilient.

As we move forward, let's carry the torch of lifelong learning and skill sharing, lighting the way for ourselves and future generations. The path won't always be easy, but the rewards—both personal and collective—are boundless. Let's embrace the journey with open minds and hearts, ready to learn, share, and innovate together.

Chapter 12:
Community and Environmental Responsibility

In this era of groundbreaking innovation and entrepreneurial spirit, the concept of responsibility towards our community and environment has never been more critical. As we delve into the realms of DIY factories and embrace new work concepts, the imperative to build sustainable practices and generate social value becomes a cornerstone of our mission. This chapter explores how integrating environmental consciousness into the fabric of our operations not only mitigates the environmental impact but also propels us towards a more sustainable future. It's about transcending the traditional metrics of success to include the well-being of our planet and its inhabitants. Through fostering a culture that values sustainability as much as innovation, we can spearhead a movement that not only revolutionizes how we work but also ensures that our endeavors contribute positively to the world around us. In essence, embracing community and environmental responsibility is not just an ethical choice but a strategic imperative that enriches our work, enhances our legacy, and ensures that our journey towards innovation is both inclusive and impactful.

Building Sustainable Practices

In the heart of the movement towards a new era of work and manufacturing lies a profound commitment to building sustainable practices. It's about more than just reducing waste or conserving energy; it's about embedding sustainability into every aspect of the

DIY factory and New Work philosophy. This approach ensures that as we stride forward, we're not just creating value for ourselves and our communities but also safeguarding the planet for future generations. The key lies in innovative thinking that challenges the status quo, leveraging technology and collaborative effort to minimize environmental impact while maximizing social value. By fostering a culture that places a premium on sustainability, we take a significant step towards a model of work that is not only productive and creative but also responsible and forward-thinking. In doing so, we invite a future where business success and environmental stewardship are not at odds but are complementary aspects of a smarter, more resilient approach to modern work.

Environmental Impact of DIY Factories The transformative wave of the DIY factory movement is not just remodeling the work culture and fostering innovation, but it's also setting new benchmarks for environmental responsibility. At the heart of the DIY ethos is a consciousness about the footprint we leave on our planet, a concern that has been somewhat sidelined in traditional manufacturing processes. As we delve into the environmental impact of DIY factories, it's crucial to recognize how these innovative spaces are not only breeding grounds for creativity and collaboration but also champions of sustainability.

The very foundation of the DIY factory model challenges the conventional norms of consumption and production. Where traditional factories may rely heavily on mass production, often resulting in significant waste and environmental degradation, DIY factories advocate for and practice a more sustainable ethos. This includes the reutilization of materials, emphasizing repair over replacement, and a profound dedication to recycling and upcycling practices. Such endeavors significantly reduce waste, lower carbon footprints, and cultivate an environment of conscientious production and consumption.

Energy consumption is another critical area where DIY factories are pioneering change. By harnessing renewable energy sources and employing energy-efficient practices, these spaces are setting a new standard for minimizing environmental impact. The use of solar panels, wind turbines, and even geothermal energy sources not only powers these innovative factories but also inspires a shift toward sustainable energy use in the broader manufacturing sector.

Water usage, often a significant environmental concern in traditional manufacturing processes, is approached with a thoughtful conservation mindset in DIY factories. Implementing water-saving techniques, such as rainwater harvesting and the reuse of greywater, these factories demonstrate how industry can operate within the means of our natural water cycles, significantly reducing their ecological footprint.

Beyond physical resources, the DIY factory movement is also recalibrating the ecological impact of logistics and transportation. By localizing production, these factories not only bring makers and consumers closer together but also drastically reduce the emissions associated with long-distance transport of goods. This localization fosters a sense of community around production, emphasizing the environmental benefits of local sourcing and sales.

Moreover, the very ethos of DIY factories encourages innovation in materials, pushing for the development and use of sustainable, non-toxic materials. This move away from hazardous chemicals and materials not only improves the safety and well-being of workers but also significantly mitigates the environmental hazards associated with traditional manufacturing processes.

The integration of technology within DIY factories presents another avenue for environmental innovation. By utilizing advanced manufacturing technologies such as 3D printing, these factories can produce goods on-demand, reducing waste associated with overproduction and unsold inventory. Additionally, these technologies allow for precise usage of materials, further minimizing waste.

Community engagement and education play a vital role in amplifying the environmental impact of DIY factories. Through workshops, seminars, and open days, these spaces disseminate knowledge on sustainable practices and the importance of environmental stewardship. This not only cultivates a more informed and conscientious community but also inspires collective action towards a more sustainable future.

Another significant environmental advantage is the tendency of DIY factories to rejuvenate and repurpose derelict industrial spaces. Rather than contributing to urban sprawl by constructing new buildings, many DIY factories breathe new life into old structures, preserving architectural heritage while minimizing the environmental impact of new constructions.

Also, the emphasis on circular economy principles within DIY factories drives a more sustainable approach to resource use and waste management. By designing products for longevity, repairability, and recyclability, these factories challenge the disposable culture endemic in traditional manufacturing, promoting sustainability at every level of the production cycle.

The impact of DIY factories extends even to biodiversity. Environmental consciousness in material sourcing and waste management practices helps in reducing the human footprint on natural habitats, thus protecting biodiversity. This mindful approach to production showcases how industrial activities can coexist harmoniously with our planet's ecological balance.

In addressing global challenges such as climate change, DIY factories stand as beacons of hope and action. Their commitment to reducing greenhouse gas emissions through sustainable practices and renewable energy use demonstrates the potential for industry to contribute positively to the fight against climate change.

However, the journey towards fully sustainable practices is an ongoing one. Continuous innovation, experimentation, and a commitment to improvement are essential in further minimizing the environmental impact of DIY factories. This includes exploring new sustainable materials, refining energy and water use practices, and continually engaging with the community to foster an inclusive, environmentally-conscious culture.

As we look towards the future, the environmental impact of DIY factories offers a compelling vision of what is possible when innovation, community, and environmental stewardship converge. Through their actions and advocacy, these factories are not only redefining the landscape of manufacturing but also leading the charge toward a more sustainable and equitable world.

In conclusion, the environmental impact of DIY factories is profound and multifaceted. By embodying principles of sustainability, community, and innovation, these spaces challenge us to rethink our relationship with the environment and with each other. They offer not just a blueprint for a more sustainable approach to manufacturing but also a source of inspiration for anyone seeking to make a positive impact on our planet. The DIY factory movement, therefore, is not just about creating goods; it's about cultivating good—an endeavor that requires the collective effort, imagination, and dedication of us all.

Generating Social Value through New Work

The landscape of work is undergoing a seismic shift, challenging the traditional paradigms and embracing a future where generating social value is at the forefront. This transformation, rooted in the principles of New Work, is not just about enhancing productivity or fostering innovation; it's about redefining the very essence of what work means in our lives and communities. As we delve into this new era, the imperative to create meaningful, socially beneficial work becomes paramount.

At the heart of New Work is the belief that every individual has the potential to contribute uniquely and significantly to the broader society. This is not just an idealistic vision but a practical approach to revitalizing our work environments and communities. By leveraging the synergies between innovative work practices and community engagement, we can create ecosystems that not only drive economic growth but also enhance social well-being.

New Work initiatives are characterized by their emphasis on autonomy, mastery, and purpose. These principles, when applied within the framework of DIY factories and similar environments, have the power to transform mundane tasks into opportunities for creative expression and personal fulfillment. More importantly, they provide a conduit for addressing societal challenges through grassroots innovation and collaborative problem-solving.

Consider, for example, the burgeoning movement of makerspaces and DIY factories that prioritize sustainability and community empowerment. These are not just spaces for individual creativity but hubs for social innovation, where the collective intelligence of the community is harnessed to develop solutions that benefit both the environment and the populace. Here, the convergence of new technology, traditional craftsmanship, and community engagement paves the way for a more sustainable and equitable society.

The adoption of flat hierarchies and open communication further amplifies the social impact of New Work. By dismantling the barriers that traditionally siloed knowledge and restricted collaboration, these structures foster a culture of shared learning and collective responsibility. This democratization of the workplace not only empowers individuals but also encourages a sense of ownership and commitment to the organization's broader social missions.

Embedded within the DNA of New Work is the principle of lifelong learning. In a rapidly changing world, the ability to adapt and grow is critical—not just for individual success but for societal progress. New Work environments serve as catalysts for continuous education, where learning is integrated into the fabric of daily activities and individuals are encouraged to expand their horizons beyond their immediate roles.

This emphasis on learning extends beyond the boundaries of the workplace into the broader community. By fostering partnerships with educational institutions, non-profits, and other community organizations, New Work initiatives can create expansive networks of learning and support. These collaborations strengthen community ties, foster a culture of mutual support, and unlock new opportunities for social and economic advancement.

The role of technology in New Work cannot be overstated. Far from rendering human skills obsolete, emerging technologies in the realm of AI, robotics, and digital fabrication are enhancing human creativity and capacity. When harnessed responsibly, these technologies can address pressing societal needs, from healthcare to environmental sustainability, elevating the role of work from mere employment to a powerful engine for social change.

Yet, the journey towards generating social value through New Work is not without challenges. Issues of access and equity loom large, as do questions about the sustainability of such models in the face of global economic pressures. Herein lies the importance of building resilient communities that can adapt to and thrive amidst such challenges. It is about creating ecosystems that value diversity, promote inclusivity, and prioritize the well-being of all members.

Implementing New Work practices requires visionary leadership and a commitment to cultural transformation. Leaders in this new era are not just managers or supervisors; they are facilitators and champions of change. They recognize the intrinsic motivation of their teams and foster an environment where innovation, collaboration, and social responsibility are part of the everyday ethos.

As we look to the future, the potential of New Work to generate social value is boundless. From revitalizing rundown urban areas through community-driven innovation hubs to addressing global challenges through collaborative research and development, the possibilities are as diverse as they are impactful. What is required is a collective commitment to this vision and the courage to reimagine the role of work in our lives and societies.

Ultimately, generating social value through New Work is about more than just enhancing productivity or fostering innovation; it's about creating work that is deeply fulfilling, socially responsible, and inherently valuable. It's about building a future where work not only supports our livelihoods but enriches our communities and nurtures our planet.

In closing, we stand at the cusp of a new era—an era where the distinction between work and social contribution becomes increasingly blurred. As professionals, entrepreneurs, and enthusiasts, we have the opportunity, and indeed the responsibility, to shape this future. By embracing the principles of New Work and integrating them into our professional environments, we can create not just a more innovative and productive workforce, but a more equitable, sustainable, and vibrant world for generations to come.

The journey ahead is both exciting and daunting. Yet, by working together, learning from each other, and staying committed to the broader vision, we can turn the ideals of New Work into a tangible reality. The promise of generating social value through New Work is not just an aspiration but a pathway to a more humane, resilient, and flourishing society.

Chapter 13:
Future Trends in Work and Manufacturing

As we pivot toward the future, a confluence of breakthrough technologies and visionary concepts stands to redefine the landscape of work and manufacturing. The integration of virtual and augmented reality into the workflow is not just an embellishment but a transformative approach to design, training, and problem-solving, enabling a level of precision and immersion previously unattainable. Moreover, the advent of digital twins, which create virtual replicas of physical systems, will revolutionize product development and lifecycle management, offering unprecedented insights and optimization opportunities. In this broader spectrum, the shift toward global networks and decentralized production heralds a new era of collaboration and innovation. These networks promise to dismantle traditional barriers, allowing ideas and products to flow freely across borders, fostering a truly global marketplace of innovation. This chapter delves into these transformative trends, sketching a future where work and manufacturing are not just about efficiency and output but about creativity, collaboration, and sustainability. In this future, the fusion of technology and human ingenuity paves the way for a manufacturing renaissance that is both inclusive and revolutionary, challenging professionals, entrepreneurs, and enthusiasts to rethink conventional paradigms and embrace the dynamic landscape of possibilities that lie ahead.

The Role of Virtual and Augmented Reality

In the enthralling journey towards the future of work and manufacturing, the advent of Virtual Reality (VR) and Augmented Reality (AR) technologies stands as a beacon of innovation, transforming how we conceive, design, and produce. These technologies not only dissolve the boundaries between the physical and digital realms but also catalyze a deeper level of engagement and creativity among professionals. Imagine designing a complex machine part from the comfort of your office, then testing its feasibility in a virtual space that mimics real-world physics – this is the game-changing potential of VR. AR, on the other hand, overlays digital information onto the physical environment, offering real-time assistance and enriching the learning and production processes. These tools are not just about enhancing efficiency; they're about enlivening the creative spirit, enabling makers and thinkers to explore uncharted territories without the constraints of the physical world. As we stand on the precipice of this new era, embracing VR and AR in our workspaces promises to lead us into a future where our manufacturing capabilities and creative aspirations are boundless.

Digital Twins in Production and Design

The transformation of manufacturing and design processes is upon us, with digital twins leading the charge. This innovative technology, at its core, involves creating a virtual replica of a physical product, process, or system. This virtual model can run simulations, study performance issues, and generate improvements, all without having to make changes to the physical counterpart. The concept may sound futuristic, but it's being implemented by forward-thinking professionals and entrepreneurs today, inspiring a new age of innovation and efficiency.

Imagine the implications for production environments. A digital twin allows for the experimentation and testing of new designs, configurations, or processes in a virtual space. This means significantly reduced material waste and a leap in the speed of iteration. Designers and engineers can tweak and optimize products and processes in real-time, pushing the boundaries of what is possible in production and design without the traditional constraints of physical prototyping.

In the sphere of design, digital twins offer an unparalleled advantage. They enable designers to visualize not just the aesthetic aspects of a product but to simulate how it will interact with the real world and other systems. It's a holistic approach that integrates form and function, ensuring that the final product is both beautiful and practical. With this technology, designers aren't just creating objects; they're crafting experiences and interactions that mesh seamlessly with the physical world.

The potential benefits of digital twins in production and design are not confined to efficiency and innovation alone. They also herald significant cost reductions. The ability to test and refine in a virtual space means fewer physical prototypes, less material usage, and a shorter time to market. Moreover, when a product is updated or a new version is released, the digital twin can be adjusted accordingly, meaning that the design and production process doesn't have to start from scratch each time.

Yet, implementing digital twins isn't merely about adopting new software or technology. It demands a shift in mindset and culture within organizations. Teams must embrace collaboration more fully, as digital twins bring together different departments—design, engineering, production—in a cohesive, integrated process. This collaborative environment is pivotal not just for leveraging digital twins but for fostering a culture of innovation and continuous improvement.

Furthermore, digital twins contribute to sustainability efforts. By optimizing production processes and reducing the need for physical prototypes, companies can significantly lower their environmental impact. In a world increasingly concerned with sustainability, adopting digital twins can not only improve efficiency and reduce costs but also bolster a company's reputation and market appeal.

Adapting to this technology also means considering the ethical implications. Data security becomes paramount as digital twins often involve processing and storing vast amounts of proprietary and sensitive information. Ensuring that this data is protected against theft and misuse is a critical concern that companies must address as part of their digital twin strategy.

The educational aspect shouldn't be overlooked, either. For industries to fully leverage digital twins, there needs to be a workforce skilled in their use and development. This means integrating digital twin technology into STEM programs and ensuring that ongoing professional development opportunities are available for current employees. It's an investment in the future, enabling a smooth transition to more innovative and efficient practices.

Looking ahead, the evolution of digital twins promises even greater integration with emerging technologies. The pairing of digital twins with AI and machine learning can lead to "smart" systems that can predict failures, suggest optimizations, and even automate aspects of the design and production process. The potential for innovation is boundless, with these technologies driving the industries towards more personalized, adaptive, and resilient manufacturing processes.

Implementing digital twins requires foresight and a willingness to invest in new technologies. However, the companies that do so are positioning themselves as leaders in the future of manufacturing and design. They'll be at the forefront of a movement that values efficiency, innovation, and sustainability.

For entrepreneurs and forward-thinking professionals, the message is clear: embracing digital twins is not just about staying ahead of the technological curve. It's about leading the charge towards a more innovative, efficient, and sustainable future. It's an opportunity to redefine not just how products are designed and produced, but how industries function at their core.

As we forge ahead, the integration of digital twins in production and design will undoubtedly face challenges—from technological hurdles to resistance to change. Yet, with the right mindset, approach, and leadership, these obstacles can be overcome. The future belongs to those who are bold enough to embrace new possibilities, daring to imagine and create a world where digital and physical realms merge seamlessly, enhancing every facet of production and design.

In conclusion, the adoption of digital twins signifies a monumental shift in production and design processes. It's a shift that demands not just technological adoption but cultural transformation within organizations. By fostering collaboration, driving innovation, and committing to continuous improvement, companies can harness the full potential of digital twins. The future is bright for those who choose to embrace this transformative technology, paving the way for a more innovative, efficient, and sustainable world.

Global Networks and Decentralized Production

In the grand tapestry of the future of work and manufacturing, the concepts of global networks and decentralized production stand out as both revolutionary and essential. As we navigate the complexities of the 21st century, the allure of traditional centralized production fades, making way for more agile, resilient, and democratized models. This paradigm shift is not just a fleeting trend but a robust response to the evolving demands of global markets, consumer expectations, and the relentless pace of technological advancement.

At its core, decentralized production is about dispersing the processes of creating goods across various locations, often leveraging local resources, talents, and capacities. This model contrasts starkly with the monolithic factories of the past, where production was concentrated in massive facilities, often in a single geographical area. Today, we're witnessing a remarkable transformation, where small, nimble units of production, spread across the globe, collaborate through sophisticated digital networks. This configuration offers unparalleled flexibility, enabling companies to respond swiftly to changes in demand and to innovate at a rapid pace.

One might wonder, what's driving this shift towards decentralization? The answer lies in the advent of groundbreaking technologies such as 3D printing, the Internet of Things (IoT), and cloud computing, among others. These innovations have made it feasible and cost-effective to establish small-scale production units that can operate with the precision and efficiency once reserved for larger factories. Moreover, the digitization of supply chains and the ability to share real-time data across continents enable these dispersed units to function as a cohesive whole. This is the power of global networks in action, transforming the landscape of manufacturing and work.

Consider the implications of such a model on the concept of work itself. Decentralized production empowers individuals and small teams, offering them autonomy and the opportunity to harness their creativity and local insights. This approach aligns perfectly with the ethos of the DIY factory and new work, where the value of individual initiative, collaboration, and innovation is paramount. It's a model that not only democratizes production but also enriches job quality, leading to more fulfilling and engaging work environments.

Moreover, global networks facilitate an unprecedented level of collaboration across borders. They allow for the seamless exchange of ideas, best practices, and innovations, further accelerating the pace of development and improvement. This ecosystem nurtures a culture of continuous learning and adaptation, critical in an era where change is the only constant.

Yet, embracing global networks and decentralized production is not without its challenges. Organizations must navigate issues related to quality control, intellectual property protection, and the coordination of highly distributed teams. However, these hurdles are not insurmountable. With the right frameworks and a commitment to fostering open, transparent communication, companies can harness the full potential of this model.

Furthermore, decentralized production has profound implications for sustainability and environmental responsibility. By producing goods closer to where they are needed, we can significantly reduce the carbon footprint associated with logistics and transportation. Additionally, the ability to tailor production to local needs helps minimize waste, contributing to a more sustainable and ethical manufacturing model.

The social implications are equally profound. Decentralized production can drive economic development in regions previously marginalized in the global economy. By bringing production closer to diverse communities, we open up new opportunities for employment, skill development, and technological empowerment.

As we look to the future, the role of education and skill development in supporting global networks and decentralized production becomes increasingly critical. Preparing the next generation of workers requires a shift in focus towards skills such as digital literacy, collaboration, and creative problem-solving. It also calls for a reimagining of learning environments to reflect the decentralized, networked nature of the future workplace.

In conclusion, the trajectory towards global networks and decentralized production is clear. This model offers a robust answer to the demands of the modern world, promising not only efficiency and innovation but also greater equity and sustainability. The journey requires vision, courage, and collaboration, but the rewards—a more vibrant, inclusive, and resilient economy—are well worth the effort.

As we embrace this shifting paradigm, we're not just redefining manufacturing; we're reimagining what's possible in work and society. It's a moment ripe with potential, an opportunity to reshape our world in the image of our highest ideals. For forward-thinking professionals, entrepreneurs, and enthusiasts, the message is clear: the future is decentralized, networked, and profoundly human. Let's seize the opportunity to build it together.

In essence, the rise of global networks and decentralized production marks a profound transformation in how we conceive of and engage with work. It's a shift that calls for innovation, adaptability, and an unwavering commitment to leveraging technology for the greater good. The path forward is not without its challenges, but the direction is unmistakable. As we forge ahead, let us do so with the knowledge that we are not just participants in this transformation but architects of a new and promising era.

Chapter 14:
Case Study: Implementing DIY Factory Principles in Traditional Industries

In a world where the quest for efficiency often overshadows the spirit of exploration, the narrative of traditional industries transforming through the infusion of DIY factory principles offers a compelling testament to the power of innovation. This chapter delves into a series of transformative stories from a variety of sectors, showcasing how the ethos of the DIY movement breathes new life into conventional production methods. By adopting a mindset that values creativity over conformity, companies have navigated the turbulent waters of change, emerging not just unscathed but revitalized. From the floors of age-old manufacturing units to the desks in sprawling office complexes, the principles of self-sufficiency, innovation, and community have paved the way for a revolution. These principles challenge the status quo, advocating for a workplace that's not just a site of productivity, but a canvas for creative expression and mutual growth. As we explore these transformation stories, we unearth the lessons learned and best practices that have made such shifts not just possible, but overwhelmingly positive. Through this journey, it becomes abundantly clear that implementing DIY factory principles in traditional industries isn't just about altering workflow or restructuring physical spaces; it's about cultivating a culture that celebrates autonomy, fosters collaboration, and empowers every individual to be a maker and a visionary in their own right.

Transformation Stories from Various Sectors

In the pivotal journey to reshaping traditional industries with the principles of the DIY factory model and New Work concepts, the transformation stories emerging from various sectors serve as beacons of innovation and change. From manufacturing to healthcare, education to retail, leaders and teams are breaking conventional molds to reimagine what is possible. These narratives are not just tales of technology adoption but are marked by the cultural shifts towards openness, collaboration, and empowerment across levels. Organizations are discovering the value in fostering environments where experimentation and learning are encouraged, blurring the lines between roles to facilitate a more agile response to challenges and opportunities. Through this lens, the transition stories highlight not only the adaptation of new tools and processes but also the cultivation of creative confidence among employees, driving them to think like innovators and act like entrepreneurs. This transformation transcends physical spaces and technological advancements, embedding a mindset of continuous improvement and a spirit of collaboration that propels the traditional sectors into a future where the principles of DIY factories and New Work are not just aspirations but realities lived daily.

Lessons Learned and Best Practices In the journey toward integrating DIY factory principles and New Work concepts into traditional industries, countless lessons have been learned. These are invaluable for forward-thinking professionals, entrepreneurs, and enthusiasts looking to foster an environment of innovation and creativity within their organizations. Drawing from various transformation stories, we distill key insights and best practices crucial for navigating this transformative path.

Firstly, the importance of a mindset shift cannot be overstated. Traditional industries typically operate under well-established norms that resist change. Embracing DIY factory principles and New Work concepts necessitates a fundamental rethinking of how work is approached, valued, and structured. This means breaking down hierarchical barriers, valuing every team member's input, and fostering a culture where innovation is not just welcomed but is a driving force.

Creating a space that physically and philosophically supports collaboration and innovation is also essential. Gone are the days of isolated workstations and rigid office layouts. The DIY factory model thrives in open, flexible spaces where ideas can flow freely, and prototypes can come to life. This physical transformation goes hand in hand with the cultivation of a community of innovators who share a common vision and passion for creating something new and valuable.

Moreover, the implementation of flat hierarchies and open communication channels is a cornerstone of the New Work movement that has proven to be effective in traditional sectors. These practices encourage a sense of ownership among team members, facilitate faster decision-making, and foster a deeper sense of engagement and satisfaction at work.

Adapting to a culture of continuous learning and iteration is another key lesson. In the fast-paced world of DIY factories, the ability to quickly learn from failures and pivot as needed is invaluable. This requires building a team that is not only skilled but also adaptable and eager to embrace new knowledge and techniques.

Integrating technology smartly not just for automation but as an enabler of creative and efficient work practices is crucial. From leveraging artificial intelligence for data-driven decisions to employing robotics for precision in manufacturing, technology plays a multi-faceted role in enhancing both productivity and creativity.

From an economic standpoint, shifting from consumerism to production has not only tangible benefits but also positions companies as key players in sustainable development. Adopting DIY factory models can lead to reduced waste, more efficient use of resources, and a stronger connection with local communities and economies.

Addressing legal and ethical considerations upfront is vital. Aspects such as intellectual property, data privacy, and fair labor practices need careful navigation, particularly in collaborative and open-source environments.

Education and skill development play a pivotal role in preparing the workforce for the future. Companies that invest in training and skill-sharing initiatives find themselves better equipped to navigate the challenges of New Work and the DIY factory model.

Community and environmental responsibility are not just buzzwords but are integral to the success and sustainability of the DIY/New Work model. Practices that prioritize sustainability, fair trade, and community engagement have shown to not only increase brand loyalty but also contribute to a healthier society and planet.

Embracing future trends, such as virtual and augmented reality, digital twins, and decentralized production, requires a forward-thinking mindset and willingness to experiment. These technologies offer avenues for innovation that can redefine industries.

Transformation stories highlight the importance of leading by example. Leaders who embody the principles of New Work and the DIY ethos inspire their teams, drive change, and overcome resistance through their actions and commitment.

Finding a balance between passion and pragmatism is crucial for personal transformation. While it's essential to pursue innovations with zeal, successful professionals also stay grounded in practicality, ensuring that projects are viable and sustainable in the long term.

The journey towards integrating DIY Factory and New Work principles into traditional industries is not without its challenges.

However, by embracing these lessons learned and best practices, organizations can navigate the path more effectively, turning potential hurdles into stepping stones for innovation and growth.

In conclusion, the shift towards DIY Factory models and New Work practices in traditional industries is more than a trend; it's a transformative movement that is reshaping the future of work. By learning from those who have paved the way, adopting best practices, and remaining adaptable in the face of change, companies can thrive in this new era of work culture. The journey is ongoing, and each step forward brings with it the promise of creativity, productivity, and a deeper sense of fulfillment in the professional realm.

Chapter 15:
Personal Transformation:
From Employee to Innovator

In the journey from being an employee to emerging as an innovator, there's a transformative shift that happens not just in one's professional identity, but in the very core of how one perceives work and creativity. This chapter delves into the essence of cultivating an entrepreneurial mindset that is critical for anyone aiming to navigate successfully through this change. It's about seeing beyond the confines of job descriptions and embracing a holistic view of how your work impacts the world around you. Tools for personal and professional growth are integral, serving as the compass that guides you through uncharted territories of innovation. Yet, it's balancing passion with pragmatism that truly marks the arrival into the innovator's realm. This balance ensures that while your head's in the clouds of big ideas, your feet remain firmly planted on the ground of reality, navigating the practical challenges of bringing your visions to life. It's in this balance that one finds the essence of transformative growth, paving the way from traditional employment to groundbreaking innovation.

Cultivating an Entrepreneurial Mindset

In transitioning from employee to innovator, one's mindset plays a pivotal role, acting as both the catalyst and the sustenance for the journey ahead. Cultivating an entrepreneurial mindset isn't about discarding risk-aversion overnight or making leaps without looking. It's about nurturing a persistent curiosity, honing the ability to see beyond the immediate to what might be, and developing a resilience that encourages learning from every outcome, favorable or not. This entrepreneurial thinking requires an openness to novel ideas, a commitment to continuous improvement, and an appetite for solving problems in ways that diverge from the norm. As professionals strive to embed themselves within the innovative practices of DIY factories and embrace the tenets of New Work, it becomes evident that such a mindset is not innate but meticulously crafted. It involves recalibrating one's approach to failure, viewing it not as a setback but as a potent source of insight, and fostering a culture where experimentation is not just tolerated but celebrated. By investing time in tools for personal and professional growth, individuals can steadily refine this mindset, transforming not only their work but their impact on the evolving landscape of modern industries.

Tools for Personal and Professional Growth As we delve deeper into the multifaceted world of DIY factories and New Work, it's essential to arm oneself with an arsenal of tools tailored for personal and professional development. These tools aren't just about enhancing productivity; they're about reshaping your mindset, fostering innovative thinking, and propelling you towards achieving unparalleled growth.

The first tool in this endeavor is the habit of continuous learning. In an era defined by rapid change and technological advancements, the willingness to adapt and learn new skills is paramount. This doesn't merely mean pursuing formal education or training sessions. It encompasses seeking knowledge through varied channels - be it online courses, podcasts, or books that challenge your thinking and spark curiosity.

Another indispensable tool is network building. The saying "your network is your net worth" holds a deeper truth in the contemporary workforce. Building a robust network isn't just about collecting contacts but fostering meaningful relationships with individuals who inspire you, challenge your perspectives, and offer support. This exchange isn't one-sided; it's about contributing your knowledge and skills, thus creating a vibrant ecosystem of shared growth.

Time management, in conjunction with prioritization, forms another critical tool. In the push towards achieving goals, understanding how to effectively manage your time - distinguishing the urgent from the important - can make all the difference. Tools and methodologies like the Eisenhower Matrix or Pomodoro Technique can be transformative, enabling you to navigate through tasks with heightened efficiency.

Mindfulness and self-awareness are tools often overlooked in the professional setting, yet their impact on growth and innovation cannot be underestimated. The journey towards personal and professional development is replete with challenges. Cultivating a practice of mindfulness can enhance resilience, improve decision-making, and foster creativity, enabling you to meet challenges with equanimity.

Critical thinking and problem-solving skills are irreplaceable tools in your growth toolkit. In an age where information is abundant, the ability to analyze, synthesize, and apply knowledge innovatively sets you apart. Problem-solving shouldn't be viewed as a reactive measure to challenges but as a proactive skill to anticipate and mitigate potential issues.

Embracing failure as a tool for growth is vital. The fear of failure can stifle innovation and risk-taking. Reconceptualizing failure as a stepping stone to success nurtures resilience. Each failure provides invaluable lessons, pushing you closer to your goals, provided you're willing to listen and adapt.

The ability to articulate your vision and ideas clearly is another crucial tool. Effective communication can galvanize teams, attract resources, and articulate the value of your innovations. Whether it's through storytelling, presentations, or casual conversations, honing your communication skills can significantly amplify your impact.

Adopting a growth mindset, the belief that abilities and intelligence can be developed, is transformative. It encourages you to embrace challenges, persist in the face of setbacks, and see effort as a path to mastery. This mindset fosters a love for learning and resilience, which are essential for long-term success and innovation.

Developing emotional intelligence (EQ) plays a pivotal role in personal and professional arenas. The ability to understand and manage your emotions, along with empathizing with others, can lead to more productive collaborations, healthier work environments, and more effective leadership.

Lastly, leveraging technology as a tool for growth is indispensable. In the context of DIY factories and New Work, technology isn't just a facilitator of tasks but a catalyst for innovation. Familiarizing yourself with the latest tools and platforms can enhance productivity, enable remote collaboration, and unlock new possibilities for creativity and innovation.

Equipped with these tools, you're not just surviving in this new work culture; you're poised to thrive. It's about integrating these tools into your daily practice, constantly iterating, and adapting. As you embark on this journey of personal and professional growth, remember that the path isn't linear. Each step forward, each setback navigated, and each success builds upon the last, propelling you toward a future where your potential is not just realized but exceeded.

The landscape of work is evolving, and with it, the tools we use to navigate this terrain must also adapt. Whether you're an entrepreneur looking to forge your path, a professional striving to innovate within your organization, or simply someone passionate about the intersection of creativity and productivity, these tools are your compass. By cultivating these skills, embracing a mindset of growth, and leveraging the collective knowledge and experience of your network, you set the foundation for a future that is not only successful but deeply fulfilling.

As we continue to explore the principles of DIY factories and New Work, remember that these tools for personal and professional growth are not static; they evolve as you do. The journey of growth is ongoing, a perpetual cycle of learning, adapting, and innovating. And in this journey, the most powerful tool you have is your willingness to embrace the unknown, to challenge the status quo, and to persistently pursue your vision of a more creative, innovative, and fulfilling future.

Balancing Passion and Pragmatism

In the journey from being an employee to emerging as a true innovator, a critical crossroad every visionary encounters is the delicate equilibrium between passion and pragmatism. It's a delineation of dreams and reality, where the fervor to create and disrupt meets the ground realities of economic models, market demands, and the essential structures within which innovation operates.

Passion is the driving force that propels inventors, thinkers, and makers beyond the conventional boundaries of thought and action. It's the spark that ignites the undying flame of curiosity, urging you to ask "what if" and "why not". However, as transformative as passion can be, it requires the steady hand of pragmatism to navigate the inherent challenges and constraints that come with bringing a vision to life.

Pragmatism, in this context, doesn't mean curtailing ambitions or dreaming smaller. Rather, it's a framework for practicality; it's about making informed decisions, understanding the ecosystem you're operating within, and recognizing the path that can feasibly take your passionate ideas from concept to completion. It's about balancing risk with reward, and understanding that every innovative journey is a series of calculated steps rather than a singular leap.

One of the first steps in achieving this balance is acknowledging that not every idea needs to revolutionize the world. Innovation can be incremental—a series of small, pragmatic steps that iteratively improve a process, product, or system. This approach allows for the accommodation of passionate creativity within a pragmatic framework, ensuring sustained progress without the paralyzing fear of failure.

Another aspect is resource allocation. Passion might push you to invest every available resource into your idea, but pragmatism guides you to do so wisely. It teaches the importance of budgeting not just financial, but also human and temporal resources. It encourages a lean methodology, where the focus is on achieving maximum impact with minimum investment, allowing for agility and adaptability.

Fostering a culture of feedback is similarly essential. While passionate innovators are often deeply invested in their visions, a pragmatic approach values external inputs and critiques. It involves being open to feedback, learning from user experiences, and understanding market needs. This ensures that the innovation is not just a reflection of personal passion but is aligned with broader applications and utility.

Navigating the path from passion to pragmatism also involves embracing failure as a stepping stone. While passion fears failure, pragmatism sees it as an inevitable aspect of the innovation process. It views each setback not as a defeat but as a valuable lesson, an opportunity to refine and pivot, thereby strengthening the eventual success of the endeavor.

Risk management is another cornerstone. Balancing the innovator's inclination to dive headfirst into uncharted territories with a pragmatic assessment of potential pitfalls and proactive planning can prevent catastrophic failures while still allowing for bold moves.

In this balancing act, setting realistic goals and timetables is crucial. While passion sets high aims, pragmatism breaks them down into achievable milestones. This not only ensures a steady progression towards the ultimate goal but also keeps the team motivated and focused, providing clear markers of success along the way.

The role of mentorship and collaboration cannot be overlooked. Passion drives individual visionaries, but pragmatism acknowledges the power of collective wisdom. Building a network of mentors, peers, and collaborators provides a support system that can offer practical advice, share experiences, and sometimes, offer a critical reality check.

Transitioning from an idea to a market-ready innovation also demands a deep understanding of the target audience. Passion inspires the creation, but pragmatism ensures its relevance and adoption. It involves researching the market, understanding customer pain points, and tweaking the innovation to meet those needs effectively.

Above all, balancing passion and pragmatism is an ongoing process, not a one-time achievement. It requires constant recalibration as ideas evolve, markets shift, and new challenges emerge. This dynamic process ensures that the innovator's journey is as rewarding as the destination.

Ultimately, the synthesis of passion and pragmatism is what sustains long-term innovation. It's what allows visionary ideas to take flight, grounded by the realities of execution and market viability. It is this balance that transforms passionate employees into pragmatic innovators, enabling them to not just dream of change but to effectively be the architects of it.

The journey from employee to innovator is both exhilarating and daunting, but those who learn to navigate the delicate balance between their passion for change and the pragmatism required to achieve it are the ones who truly leave a mark on the world. They are the ones who understand that while passion may fuel the journey, pragmatism ensures its success, making the audacious goals of today the tangible realities of tomorrow.

Chapter 16:
Leadership in the Era of DIY
Factory and New Work

In a world where traditional hierarchical models are being challenged and reshaped, leaders in the era of DIY Factory and New Work are faced with a unique set of challenges and opportunities. To thrive, they must cultivate an environment that empowers individuals, fosters collaboration, and embraces change as a constant. This calls for a leadership style that is not only visionary but also deeply empathetic, one that recognizes the intrinsic value in every team member and understands that innovation is as much about fostering diversity of thought as it is about technology. By leading by example, these pioneers inspire creativity and a can-do spirit, encouraging their teams to experiment and take risks in a safe and supportive environment. Moreover, navigating the uncertainties of this new era requires a resilience and adaptability grounded in a clear vision and a commitment to continuous learning and improvement. This chapter delves into strategies for empowering teams, facilitating collaborative innovation, and steering through the waters of change with confidence and agility, thereby ensuring that the organization remains at the forefront of the New Work movement. Embracing these principles, leaders can not only drive their organizations towards unprecedented growth but also contribute to shaping a future where work is more meaningful, inclusive, and aligned with personal values and societal needs.

Leading by Example: Inspiring Innovation and Creativity

In the dynamic confluence of the DIY Factory and New Work paradigms, the quintessence of leadership transforms. No longer about mere direction and delegation, it becomes an art form of inspiring innovation and creativity. At its core, this leadership navigates the nuanced balance between giving autonomy and providing guidance, creating an environment where every team member feels empowered to challenge the status quo and bring forth groundbreaking ideas. It's about demonstrating a commitment to innovation not just in words but in actions, making bold decisions that encourage risk-taking, and celebrating both successes and failures as essential parts of the creative process. Leaders in this new era serve not only as visionaries but as architects of a culture that breathes experimentation, where each individual's unique talents and perspectives converge to spark continuous evolution and reimagination of what's possible. By embodying the principles they champion, leaders become the catalysts for a transformative movement, where creativity flourishes, boundaries are pushed, and the future of work is actively sculpted with intention, thoughtfulness, and an unwavering belief in human potential.

Empowering Teams and Facilitating Collaboration As we venture further into the exploration of the DIY factory and New Work ethos, it's crucial to dive deep into the elements that are the lifeblood of innovation and productivity within these frameworks: empowering teams and facilitating collaboration. These principles stand as pillars, supporting an infrastructure that thrives on creativity, autonomy, and collaborative success. In this environment, every team member is seen not just as a cog in the machine but as a vital contributor with unique insights and skills.

The essence of empowering teams lies in recognizing the intrinsic motivation of individuals and harnessing it towards collective goals. This approach differs markedly from traditional hierarchies, where commands flow downward, often stifling creativity and initiative. In contrast, a flat organizational structure, typical in DIY factories and New Work cultures, promotes open communication and shared leadership. It invites a democratic way of decision-making and problem-solving, ensuring that everyone's voice is heard and valued.

Facilitating collaboration in such an environment means creating spaces and opportunities for individuals to come together, share ideas, and synergize their strengths. This often involves designing collaborative workspaces that inspire creativity and interaction, equipping teams with tools and technologies that enhance productivity, and fostering a community of innovators who support and challenge each other.

One might wonder how such ideals are practically implemented. The answer often lies in simple yet impactful practices: regular team meetings structured around open dialogue rather than one-way reports, brainstorming sessions where unconventional ideas are encouraged, and project management approaches that allow for flexibility and adaptability. These methods ensure that teams remain agile and responsive, able to pivot as challenges arise and opportunities present themselves.

Moreover, empowering teams also means investing in their growth and development. A culture that values continuous learning and skill-sharing not only keeps the team updated with the latest trends and technologies but also fosters a sense of personal investment in each member. When individuals see their growth trajectories aligned with the organization's vision, their engagement and commitment skyrocket, driving innovation and productivity.

Leadership, within such structures, transforms from a role of authority to one of mentorship and facilitation. Leaders in DIY factories and New Work environments serve as catalysts for collaboration, offering guidance and support while stepping back to allow team members to lead initiatives and projects. This shift not only empowers teams but also cultivates a new generation of leaders who are well-versed in the principles of collaboration and empowerment.

The benefits of such an approach are manifold. Teams that feel empowered and are encouraged to collaborate bring about a higher level of innovation, as diverse perspectives come together to solve problems and explore new ideas. The sense of ownership and accountability that comes from such empowerment leads to higher satisfaction and lower turnover, as team members deeply connect with their work and see their role in the organization's success.

Yet, the journey towards truly empowering teams and facilitating collaboration is not without its challenges. It requires a deep-seated shift in mindset from both leaders and team members, moving away from the traditional notions of work and authority towards a culture of trust, respect, and mutual support. It demands an ongoing commitment to transparency and communication, ensuring that every team member feels valued and understood.

In practical terms, this often means revisiting and revising existing policies and practices – from how decisions are made, how projects are managed, to how success is measured and celebrated. It invites organizations to experiment with new tools and methodologies, constantly seeking out ways to enhance collaboration and empower teams more effectively.

Success stories from both the DIY factory movements and New Work initiatives show that such efforts are not only worthwhile but essential in today's rapidly changing work environment. Companies that have embraced these principles report not just higher levels of innovation and productivity but also a stronger sense of community and shared purpose among their team members.

As we look ahead, the need for empowering teams and facilitating collaboration becomes even more pronounced. The future of work is inherently collaborative, driven by complex challenges that demand diverse skills and perspectives. In this context, organizations that cultivate an environment where teams are empowered and collaboration is facilitated stand out, ready to navigate the uncertainties of the future with agility and resilience.

In conclusion, the journey towards empowering teams and facilitating collaboration is not a linear path but a continuous cycle of learning, adapting, and growing. It requires commitment, patience, and an unwavering belief in the potential of each individual. For those willing to embrace these principles, the rewards are immeasurable – not just in terms of business success but in fostering a work culture that is vibrant, innovative, and deeply fulfilling.

As we move forward, let's carry these insights with us, applying them in our endeavors, whether in setting up a DIY factory, transforming existing work environments, or simply seeking to cultivate a more innovative and collaborative culture. The future of work is not a distant dream but a present opportunity, beckoning us to embrace empowerment and collaboration as the keys to unlocking unprecedented levels of creativity and progress.

Navigating Change and Uncertainty

In an era that demands both rapid innovation and adaptability, leaders find themselves at the helm of navigating through a sea of change and uncertainty. The convergence of DIY factory principles with New Work concepts presents a unique set of challenges and opportunities. Leadership, in this context, requires a blend of vision, resilience, and an unparalleled commitment to fostering cultures that thrive on change rather than fear it.

Change, while inherently uncomfortable, is the bedrock of innovation. The leaders who excel in today's fast-paced environment are those who not only accept change but also champion it. They understand that navigating through uncertainty is not about having all the answers but about asking the right questions and being willing to experiment and learn from those experiments.

The concept of the DIY factory, steeped in the ethos of innovation, collaboration, and sustainability, requires a leadership style that is both hands-on and inspiring. Leaders must cultivate environments where every team member feels empowered to bring their ideas to life, no matter how unconventional. This approach does not just apply to product development but extends to every facet of operations, from workflow design to team structure.

New Work theories further emphasize the importance of autonomy, purpose, and mastery in the workplace. Leaders are tasked with creating organizational cultures that align with these values, ensuring that team members find meaningful engagement in their work. This alignment isn't just beneficial for employee satisfaction; it's critical for attracting and retaining top talent in an increasingly competitive landscape.

One of the pivotal strategies in navigating change and uncertainty is fostering a culture of continuous learning and adaptation. This involves not just upskilling but also creating mechanisms for constant feedback, reflection, and iterative improvement. Leaders must champion a mindset where failure is not a setback but an integral part of the learning process.

Open communication stands as a cornerstone in this evolving work paradigm. Transparency about changes, challenges, and the organization's strategic direction helps build trust and resilience within the team. Leaders must ensure that communication channels are not just open but also genuinely reciprocal, allowing ideas and concerns to flow freely from all levels of the organization.

Adaptability in leadership also involves rethinking traditional hierarchies and embracing more fluid organizational structures. In the spirit of the DIY factory, creativity and innovation often stem from cross-disciplinary collaboration. Leaders should encourage a culture where silos are dismantled in favor of networks of teams that come together around projects and goals, fostering a sense of unity and shared purpose.

The uncertain landscape that leaders navigate today is further complicated by the rapid pace of technological change. Embracing technology as an enabler rather than a disruptor is key. This entails staying informed about emerging trends, understanding their implications for the industry, and being proactive in integrating these innovations in a way that enhances, rather than overwhelms, human potential.

In times of change, vision becomes more critical than ever. Leaders must articulate a clear and compelling vision that serves as a north star for the organization. This vision should not only outline what the organization aspires to achieve but also why it matters. It's the 'why' that often provides guidance and motivation during times of uncertainty.

Equally important is the ability to translate this vision into actionable strategy. This requires a balance of big-picture thinking and attention to detail, ensuring that every team member understands their role in the bigger mission and feels equipped to contribute meaningfully.

Empathy and emotional intelligence play a significant role in leading through change. Understanding the human side of uncertainty—recognizing the fears, hopes, and motivations of team members—is essential in supporting them through the transition. Leaders must be adept at listening, offering support, and, when necessary, providing the push to move out of comfort zones.

Building resilience within the organization is another crucial aspect of navigating change. This involves not just financial or operational resilience but emotional and psychological resilience as well. Celebrating successes, learning from failures, and always moving forward with a sense of purpose and optimism are vital in building a resilient culture.

Leadership in the era of DIY factories and New Work is as much about personal transformation as it is about organizational change. Leaders themselves must embody the principles of adaptability, continuous learning, and innovation. This often means stepping out of traditional leadership roles and being willing to lead by example, whether that's by rolling up their sleeves to work alongside the team or by championing new ideas and approaches.

Navigating change and uncertainty is not a straightforward path. It requires a blend of strategic thinking, empathy, resilience, and the courage to lead by example. The rewards, however, are worth the journey—creating organizations that are not just surviving but thriving in the face of change, and fostering work environments that are innovative, inclusive, and genuinely fulfilling.

As leaders in this dynamic era, embracing the complexities of change and uncertainty is not just a responsibility but an opportunity—an opportunity to redefine what leadership looks like, to inspire teams to reach new heights, and to shape the future of work in ways we can only begin to imagine.

Chapter 17:
The Future of Work-Life Balance

In a world where the lines between work and life blur, the future of work-life balance becomes a canvas for innovation and a reflection of our values. As we navigate through the evolution of work practices, empowered by the DIY factory ethos and propelled by the principles of New Work, a transformative perspective emerges. It challenges the traditional dichotomies of work and leisure, urging us to redefine success and productivity in terms not just of output, but of fulfillment and happiness. This chapter delves into the core of how integrating personal passions with professional goals isn't a far-fetched dream but a tangible reality within reach. It uncovers the pivotal role of flexibility and autonomy, not as mere buzzwords, but as fundamental pillars that support a balanced, vibrant life. In the unfolding era, work-life balance is reimagined as a dynamic equilibrium, where the pursuit of innovation and the quest for a fulfilling life converge, creating a landscape ripe with opportunities for growth, creativity, and well-being.

Redefining Success and Productivity

In the ever-evolving landscape of work, the traditional metrics of success and productivity are undergoing a significant transformation. We're moving away from a world where long hours and constant availability are badges of honor, toward a paradigm where the quality of output, creativity, and well-being take center stage. This shift calls for a reevaluation of what it means to be successful and productive in both professional and personal realms. It's about integrating our passions with our professional goals, and recognizing that true productivity encompasses not just what we achieve at work, but how we enrich our lives and those around us. By embracing flexibility, autonomy, and a holistic view of success, we're not just redefining productivity—we're setting the stage for a more fulfilling and balanced approach to work and life.

Integrating Personal Passions with Professional Goals In today's rapidly evolving work environment, bridging the gap between personal passions and professional goals is not just a luxury; it has become a necessity for fostering innovation, creativity, and fulfillment. This chapter delves into how forward-thinking professionals, entrepreneurs, and enthusiasts can integrate their personal passions with their professional ambitions to unlock a new era of productivity and satisfaction.

At the heart of integrating personal passions with professional goals is the understanding that work doesn't have to be a monotonous, life-draining experience. Instead, work can be a vibrant, life-affirming journey that not only meets economic needs but also fulfills deeper, more personal aspirations. This shift requires a fundamental rethinking of what it means to work and what professional success looks like.

The convergence of personal passions and professional goals leads to a more engaged and motivated individual. When people work on projects or within companies that align with their personal interests and values, they naturally bring a higher level of energy and commitment to their roles. This alignment not only enhances job satisfaction but also drives innovation, as passionate individuals are more likely to challenge the status quo and think outside the box.

However, integrating personal passions with professional goals is not a one-size-fits-all solution. It requires a deep understanding of one's own interests, strengths, and values. Self-reflection becomes a critical tool in this journey, enabling individuals to uncover what truly motivates them and how these drivers can be aligned with their professional life.

For entrepreneurs and leaders striving to infuse new work principles into their organizations, fostering an environment where team members can pursue passion-driven projects is essential. This means creating a culture that encourages experimentation, values diverse interests, and supports continuous learning. Such an environment not only attracts top talent but also retains them by offering a deeply rewarding work experience.

One practical step towards integrating personal passions with professional goals is the development of a personal mission statement. This statement serves as a compass that guides individuals in making career decisions that align with their personal passions. It helps in identifying opportunities and roles that resonate with one's values and interests, making work more meaningful and fulfilling.

Moreover, the rise of the gig economy and flexible working arrangements has made it easier for individuals to tailor their work life in a way that better suits their personal preferences and lifestyle. Whether it's freelancing, part-time work, or remote positions, there are now more options available to pursue professional endeavors that align with personal passions.

Networking plays a pivotal role in integrating personal passions with professional goals. Engaging with like-minded individuals, attending industry conferences, and participating in online communities can open doors to opportunities that resonate with one's personal interests. These connections can lead to collaborative projects, mentorship relationships, and even new career paths that one might not have considered otherwise.

Technology also plays a crucial role in merging passions with professional pursuits. With access to online learning platforms, individuals can easily acquire new skills and knowledge that align with both their interests and career aspirations. This continuous learning mindset is essential in today's fast-paced work environment, enabling individuals to stay relevant and competitive.

Incorporating personal passions into professional goals is not without its challenges. It requires a balance between pragmatism and passion, as economic realities and market demands cannot be ignored. However, by embracing flexible work models, leveraging technology, and continuously adapting, individuals can navigate these challenges successfully.

Ultimately, the integration of personal passions with professional goals leads to a more holistic approach to work, where success is measured not just by economic gains but by personal fulfillment and societal contribution as well. This approach not only benefits individuals by making their work life more satisfying but also propels organizations forward by fostering a culture of passionate, engaged, and innovative employees.

Case studies of individuals and organizations that have successfully integrated personal passions with professional goals can provide valuable insights and inspiration. From small startups to large corporations, there are numerous examples of how aligning personal values with professional endeavors can lead to remarkable outcomes, demonstrating that this integration is not only possible but also beneficial on multiple levels.

To move towards this integration, both individuals and organizations need to embrace a growth mindset, recognizing that personal and professional development is a continuous journey. Setting aside time for reflection, being open to new opportunities, and remaining flexible in the face of change are essential strategies for integrating personal passions with professional goals effectively.

In conclusion, integrating personal passions with professional goals is a transformative process that can lead to a more fulfilling work life and a more innovative and resilient professional environment. By taking proactive steps towards this integration, individuals and organizations alike can unlock new potentials and pave the way for a more dynamic and satisfying future of work.

The Role of Flexibility and Autonomy

In the quest for a more fulfilling work-life balance, the significance of flexibility and autonomy can't be overstated. These elements are not just buzzwords; they are foundational to the transformation we're witnessing in the modern workplace. By empowering employees with greater control over their work and the conditions under which they perform it, businesses are unlocking unprecedented levels of creativity, satisfaction, and, ultimately, productivity.

Flexibility in the context of work refers to the ability to adapt and change one's work schedule, location, and even the nature of the work itself to better fit personal commitments and preferences. This flexible approach challenges the traditional 9-to-5, office-bound paradigm, embracing instead a model that accommodates individual needs and rhythms. The rise of remote work technologies has facilitated this shift, enabling people to contribute from anywhere in the world, at times that suit them best.

Autonomy goes hand in hand with flexibility, offering individuals the power to make decisions about their work without excessive oversight. This self-direction fosters a culture of trust and respect, where creativity flourishes. Autonomy encourages experimentation and innovation, as individuals feel empowered to try new approaches without fear of failure.

The benefits of incorporating flexibility and autonomy into work environments are manifold. For starters, employees report higher job satisfaction and lower stress levels, as they're able to manage their professional and personal lives more effectively. This balance is critical in preventing burnout, a phenomenon all too common in high-pressure, rigid work settings.

From an organizational perspective, businesses that champion flexibility and autonomy often see a boost in employee retention. Talented professionals are more likely to stay with a company that respects their needs and offers them the freedom to work in ways that best suit them. This retention saves businesses significant amounts in turnover costs and helps maintain a stable, experienced workforce.

The impact on innovation should not be underestimated. When workers have the autonomy to explore and the flexibility to set their own schedules, they're more likely to think creatively and come up with innovative solutions. This environment is conducive to continuous improvement and can be a significant competitive advantage in fast-paced industries.

Moreover, embracing flexibility and autonomy can enhance a company's appeal to a broader talent pool. In today's globalized world, the best candidate for a position might not reside in the same city or even the same country as the company. By offering flexible work arrangements, businesses can attract talent from anywhere, breaking down geographical barriers and enriching their team with diverse perspectives.

However, implementing these principles is not without its challenges. For many organizations, shifting to a more flexible and autonomous work model requires a foundational change in company culture. It demands robust infrastructure to support remote work, clear communication channels, and a reevaluation of how performance is measured and rewarded.

Leadership roles also evolve in this new paradigm. Managers accustomed to overseeing work directly must learn to trust their teams and focus on outcomes rather than micromanaging processes. This shift can be uncomfortable but is essential for fostering a culture of autonomy and innovation.

Furthermore, while flexibility and autonomy offer many benefits, they're not one-size-fits-all solutions. Some employees may struggle with the lack of structure, feeling isolated or procrastinating in the absence of a traditional work environment. Organizations need to recognize these challenges and offer support, whether through training, mentorship programs, or opportunities for social interaction among remote team members.

Despite these hurdles, the trend towards greater flexibility and autonomy in the workplace is unmistakable and accelerating. As technology continues to evolve, and societal attitudes towards work shift, businesses will need to adapt to attract and retain top talent. Those that do will be well-positioned to thrive in the future of work, marked by creativity, innovation, and a deeper sense of work-life balance.

Ultimately, the move towards flexibility and autonomy is not just about enhancing productivity or innovation—it's about reshaping our very understanding of what work can and should be. It's a recognition that when individuals thrive, organizations do too. By creating environments that value personal well-being as much as professional achievements, companies can cultivate more engaged, motivated, and fulfilled workforces.

To navigate this new landscape, organizations, leaders, and employees alike must be willing to embrace change, experiment, and learn from both successes and failures. The future of work isn't just about where we work or the hours we keep but about creating meaningful, satisfying careers that integrate seamlessly with the rest of our lives. In this future, flexibility and autonomy are not just advantages; they are imperatives.

As we chart a course towards this promising horizon, it's clear that the role of flexibility and autonomy will only grow. Those who understand and harness these principles will lead the way, crafting work environments that are not only more productive and innovative but also more humane and fulfilling. In doing so, they will not only revolutionize work but also enrich lives, building a future where work serves life, not the other way around.

Chapter 18:
Scaling DIY Factories and New Work Practices

In an era where the boundaries of innovation are constantly being expanded, the models of DIY factories and New Work practices have surfaced as beacons of modern industrial evolution. These paradigms aren't mere trends; they are the foundation of a significant shift in how we perceive work, creativity, and production. The challenge now lies in scaling these models beyond their local successes to a global stage, fostering a network that transcends borders and cultural barriers. This transformation demands a strategic approach, where the integration of technology, human creativity, and sustainable business practices converges to form scalable operations. By leveraging the collective power of collaborations and partnerships, DIY factories can amplify their impact, driving the New Work movement towards a future where innovation is not only nurtured but also celebrated across communities worldwide. Embracing these practices entails pioneering a shift towards a more connected, efficient, and creative global workforce, setting a new standard for what is possible when passion meets practice on an expansive scale.

From Local Impact to Global Influence

The journey from igniting a spark of change within local confines to blazing a trail on a global scale is both exhilarating and daunting. However, when the innovative practices honed in DIY factories and New Work environments begin to scale, the potential for transformative impact knows no bounds. It's about taking the ethos of creativity, collaboration, and sustainable production and embedding it into the fabric of global work culture. This expansion not only amplifies the economic models supporting DIY factories but also propels the principles of New Work onto a worldwide stage. As these practices cross borders, they foster a network of like-minded individuals and organizations, united in the pursuit of challenging the status quo. This ripple effect can democratize innovation, making it accessible to all corners of the globe, and in turn, inspire a new generation of thinkers, creators, and innovators. The scale of impact is only limited by our collective imagination and willingness to embrace and act upon these pioneering concepts, ensuring they are not just fleeting trends but foundational pillars of a global movement towards a productive, creative, and fulfilling future.

Strategies for Scaling Up Embarking on the journey of expanding DIY factories and New Work practices requires a blend of vision, strategy, and tenacity. For forward-thinking professionals and entrepreneurs, scaling is not just about enlarging the footprint but also about deepening the impact. It's a process that demands careful planning, a collaborative spirit, and an adaptive mindset.

In the realm of scaling up, one fundamental truth stands out: growth is multifaceted. It encompasses not only the physical expansion of facilities or the numerical increase in output but also the broadening of influence and the strengthening of community ties. To navigate this complex landscape, a set of key strategies must be employed, each tailored to harness the unique potential of DIY factories and New Work principles.

First and foremost, the importance of leveraging technology cannot be overstated. In a world where digital transformation shapes every aspect of work and production, DIY factories must embrace cutting-edge tools and platforms. From AI and robotics enhancing efficiency and creativity to cloud-based collaboration tools connecting dispersed teams, technology serves as a powerful enabler for scaling efforts.

At the same time, fostering a culture of innovation within the organization is crucial. This means not only encouraging the free flow of ideas but also providing a support structure for experimentation and prototyping. A culture that celebrates risk-taking and learns from failure sets the stage for breakthroughs that can propel the organization to new heights.

Building partnerships and networks emerges as another essential strategy. Collaboration across industries and sectors can unlock new opportunities, share resources, and co-create value. By engaging with suppliers, customers, academia, and even competitors, DIY factories and New Work initiatives can amplify their impact and reach.

Customer engagement and community building also play pivotal roles in scaling up. By actively involving users and the wider community in the co-creation process, organizations can foster a sense of ownership and loyalty. This not only enhances product development through real-world feedback but also helps in building a robust support base that can accelerate growth.

An agile approach to growth is another key element. In an era marked by rapid change and uncertainty, the ability to adapt and pivot is invaluable. This means staying attuned to market trends, technological advancements, and societal shifts, and being prepared to reassess and adjust strategies accordingly.

Furthermore, a focus on sustainability and ethical practices is integral to scaling up in today's world. As organizations expand, their environmental footprint and social impact become more pronounced. Embracing sustainable practices not only mitigates negative effects but also resonates with increasingly conscientious consumers and workers, fostering long-term loyalty and trust.

In addition to these strategic considerations, the operational aspect of scaling up demands attention. Streamlining processes, optimizing supply chains, and enhancing quality control are all crucial for maintaining operational excellence while growing. These internal improvements ensure that as the organization scales, it remains efficient and responsive to customer needs.

Financial planning and management are also critical for a successful scaling journey. Securing funding, whether through venture capital, crowdfunding, or other sources, requires a compelling vision and solid business plan. Furthermore, as organizations grow, managing cash flow and investments wisely becomes even more important to ensure stability and foster continued expansion.

On the human resources front, scaling up means not only increasing headcount but also nurturing talent and preserving the organization's culture. Developing leadership within, adopting flexible work arrangements, and maintaining open lines of communication can help manage the growing pains that often accompany expansion.

Educational initiatives and skill development programs can further support scaling efforts by ensuring that both current employees and the broader community are equipped with the knowledge and skills needed for success. Investing in education reflects a commitment to continuous improvement and innovation, key drivers of growth in the New Work era.

To navigate the regulatory and legal landscapes, especially as operations cross borders, is another crucial consideration. Understanding and complying with relevant laws, standards, and cultural norms are vital to successful expansion. Strategic partnerships and local expertise can prove invaluable in navigating these complex environments.

Lastly, effective communication and storytelling are indispensable tools for scaling up. Articulating the vision, impact, and successes of the organization in a compelling manner can attract partners, customers, and talent. Sharing the journey transparently builds trust and fosters a sense of community among stakeholders, further fueling growth.

In conclusion, scaling up DIY factories and embracing New Work practices is a multifaceted endeavor that requires a careful mix of innovation, collaboration, agility, and stewardship. By adopting these strategies, organizations can not only expand their operations but also amplify their impact, paving the way for a future that values creativity, community, and sustainable growth.

The Network Effect: Collaboration Across Borders

In an age where the world is more connected than ever, the principles of Do-It-Yourself (DIY) factories and New Work are not confined to geographical limitations. This chapter explores how leveraging the network effect can propel these innovative work practices beyond local boundaries, ensuring a global impact. The collaboration across borders is not just a possibility but a necessity for scaling DIY factories and embedding New Work practices in the global workforce.

The essence of the network effect lies in the fact that the value of a network increases exponentially with each additional participant. This principle, applied to the collaboration among DIY factories and New Work practitioners, can unleash unparalleled creativity, innovation, and efficiency. By connecting with fellow innovators across the globe, individuals and organizations can tap into a wealth of knowledge, resources, and markets previously out of reach.

Crafting the Future: The DIY Factory & New Work Nexus

One of the most compelling aspects of this global collaboration is the shared learning experience. Imagine a scenario where a DIY factory in Berlin shares its blueprint for sustainable furniture with a New Work community in Seoul. In return, the Seoul community shares its advancements in ergonomic design. This cross-pollination of ideas does not just lead to improved products and processes; it fosters a global culture of innovation and mutual growth.

Technological advancements, especially in communication and collaboration tools, have made it easier than ever to connect and collaborate across continents. Platforms that support project management, file sharing, and real-time communication act as the infrastructure that supports this global network of innovators. They ensure that distance is no longer a barrier to collaboration, enabling teams to work together as if they were in the same room.

However, successful collaboration across borders requires more than just the right tools. It requires a shift in mindset. Embracing diversity and fostering an inclusive culture becomes critical. Global teams bring together diverse perspectives that, when harnessed effectively, can lead to more innovative solutions. But this diversity also presents challenges, including language barriers and cultural differences that can lead to miscommunication and misunderstanding.

To navigate these challenges, it's important for teams to develop strong communication skills and cultural sensitivity. Establishing clear communication protocols and taking the time to learn about each other's cultures can go a long way in mitigating misunderstandings and building a strong, cohesive team.

Another key aspect of successful global collaboration is building trust. Trust is the foundation of any effective team, but it can be more difficult to establish at a distance. Regular communication, transparent processes, and shared goals can help build trust among team members, even when they are spread across the globe.

Case studies of successful global collaborations in the DIY and New Work spaces illustrate the power of the network effect. For instance, a global project involving DIY enthusiasts from around the world collaborating to create an open-source solution for clean water access in developing countries showcases the potential of collaborative innovation to address real-world problems.

These collaborations also highlight the importance of a shared purpose. When individuals and organizations come together around a common goal, whether it's environmental sustainability, social justice, or technological innovation, they can achieve far more than they could alone. This shared purpose becomes the glue that holds the global network together.

The role of leadership in nurturing and guiding these global collaborations cannot be understated. Leaders must be visionaries who can see beyond local and national interests, embracing a global perspective. They must also be adept at managing diverse teams, fostering a culture of inclusivity, and navigating the complexities of global collaboration.

Moreover, scaling DIY factories and New Work practices globally offers an opportunity to rethink economic models. By leveraging the collective power of a global network, these initiatives can tap into new markets and distribution channels, potentially disrupting traditional business models and paving the way for a more sustainable and equitable global economy.

Legal and ethical considerations also come into play when collaborating across borders. Intellectual property rights, data protection laws, and ethical standards vary greatly between countries and must be navigated carefully. Clear agreements and a deep understanding of international laws and regulations are essential for successful global collaboration.

The environmental impact of scaling DIY factories and New Work practices through global collaboration is also significant. By sharing best practices for sustainability and leveraging collective purchasing power, the global network can reduce its environmental footprint and contribute to a more sustainable future.

Finally, the network effect, applied through collaboration across borders, invites us to imagine a future where work is not just about productivity, but about creating value collectively on a global scale. It offers a vision of the future where innovation, sustainability, and inclusivity are not just aspirations but realities, made possible by the power of global collaboration.

As we forge ahead, the challenge and opportunity lie in how we harness this network effect not just to scale DIY factories and New Work practices, but to reimagine and reshape our global work culture. It's a journey of transformation, grounded in collaboration, that promises to redefine what's possible when we come together across borders.

Chapter 19:
The DIY Factory and New Work
as Social Movements

In a world where the very definition of work is constantly being rewritten, the emergence of DIY factories and New Work as robust social movements marks a significant pivot towards a more inclusive, innovative, and fulfilling professional landscape. These paradigms don't just challenge the status quo; they're reshaping it, offering a blueprint for the future that values creativity, autonomy, and sustainable growth over outdated metrics of success. At the heart of this chapter, we explore how these movements aren't confined to the fringes of work culture but are advocating for transformative changes in policies and perspectives across the globe. Through the lens of social entrepreneurship and innovation, we witness a shift towards a more connected, empathetic, and adaptable workforce. This isn't about a fleeting trend, but a profound evolution in how we conceive, create, and collaborate. By embracing these principles, professionals, entrepreneurs, and enthusiasts alike are not just participating in a movement; they're leading it, paving the way for a world where work is not just a means to an end but a source of personal fulfillment and societal impact. The DNA of DIY factories and New Work philosophies is embedded in the desire for change—change that fosters a global community of practitioners determined to redefine success, not just for themselves but for future generations. This chapter is a call to action, inspiring readers to become catalysts of this transformation, advocating for work cultures where innovation thrives, policies that support flexible, passion-driven careers, and a future where the boundaries between work and personal satisfaction blur into insignificance.

Advocating for Change in Work Cultures and Policies

In the realm of work, a seismic shift is afoot, and it's underpinned by the principles of DIY factories and New Work. This transformation isn't just about altering the space we operate in; it's about redefining the very ethos that guides our professional lives. To pave the way for these innovative changes, we need to champion a cultural and policy-related renaissance that not only supports but also accelerates the adoption of more collaborative, creative, and autonomous work environments. It's about creating workplaces that aren't merely places of employment but communities of practice that foster continuous learning, experimentation, and a shared sense of purpose. This requires a bold reimagining of leadership roles, moving away from traditional hierarchies and towards models that privilege empowerment and innovation at every level. Moreover, advocating for policy changes that support flexible work arrangements, protect intellectual property while encouraging collaboration, and fund initiatives that push the boundaries of what work can be, is crucial. Embracing these changes means stepping into a future where work is not just a means to an end but a fulfilling, integral part of the human experience.

Building a Global Community of Practitioners In this era of rapid change and innovation, the evolution of work practices is not just a trend but a necessity. The fusion between DIY factories and New Work principles is a beacon of progress, showcasing a model that merges creativity with productivity. However, the real power of this movement lies in its communal nature. Building a global community of practitioners is not just an ambition; it's a critical strategy for sustaining and amplifying the transformative potential of these practices.

At the heart of this burgeoning community are individuals and groups across the globe who share a common vision. They see beyond the constraints of traditional work paradigms, embracing a future where work is not just a means to an end but a source of personal fulfillment and social value. The question then becomes, how do we facilitate the growth of this community, ensuring that it remains inclusive, collaborative, and impactful?

Cross-border collaboration is a cornerstone of this global community. With today's technology, the exchange of ideas, tools, and practices has never been easier. Online platforms and social media networks provide forums where practitioners can share successes and setbacks alike, learning from each other in a continuous loop of improvement. But digital interaction is only part of the equation. The real strength of the community comes from the meaningful connections that are formed - partnerships that transcend geographical boundaries and foster innovation.

Knowledge sharing is another critical aspect. Veterans in the fields of DIY factories and New Work practices bring invaluable insights and experiences to the table. By mentoring newcomers, they ensure that the community remains vibrant and that its collective knowledge base keeps expanding. This culture of mentorship and apprenticeship not only accelerates individual learning but also strengthens the community's foundations, making it more resilient to challenges.

Education plays a pivotal role in community building, not in the traditional sense of classrooms and textbooks but through hands-on workshops, webinars, and collaborative projects. These learning experiences are designed not just to impart skills but to inspire a mindset shift towards innovation, collaboration, and continuous improvement. They are a testament to the community's commitment to not just participating in the future of work but actively shaping it.

Events and summits serve as catalysts for community growth, providing platforms for face-to-face interactions, networking, and the showcasing of innovations. These gatherings, whether they're local meetups or international conferences, are not just about celebrating achievements. They are about finding common ground, forging new collaborations, and setting collective goals for the future.

At the same time, celebrating diversity within this global community is crucial. The strength of the community lies in its diversity - of backgrounds, skills, perspectives, and experiences. This diversity fuels creativity and drives innovation, enabling the community to tackle challenges from multiple angles and come up with holistic solutions.

The community's ethos of open-source sharing and collaboration is revolutionary. By openly sharing resources, tools, and processes, practitioners embody the principles of generosity and collective progress. This open-source mentality not only accelerates innovation but also democratizes access to tools and knowledge, breaking down barriers to entry for individuals and groups worldwide.

Advocacy and policy influence are also important roles for this community. By uniting voices from around the world, the community can advocate for changes in work cultures, policies, and practices that align with New Work principles. This collective advocacy has the power to influence not just businesses but governments and international bodies, contributing to systemic change.

However, building this global community is not without its challenges. Differences in language, culture, and access to resources can create barriers to full participation. It's imperative that the community remains conscious of these challenges and actively works to overcome them. This includes ensuring that resources are accessible in multiple languages, fostering cultural exchange and understanding, and working to bridge the digital divide.

At its core, the movement towards DIY factories and New Work is about more than just changing how we work; it's about changing how we live. It's about creating a future where work is aligned with our deepest values and aspirations, where creativity and collaboration are not just encouraged but embedded in the fabric of our professional lives. Building a global community of practitioners is essential to realizing this vision, for it is together that we can push boundaries, innovate, and create a more fulfilling and sustainable future for all.

In conclusion, as we continue on this journey, it's important to remember that building a global community of practitioners is an ongoing process, one that requires patience, commitment, and a willingness to learn and adapt. It's about celebrating our achievements while staying humble and curious, always looking for new ways to grow and improve. The future of work is not a destination but a journey, and it's a journey that we are on together. The power of a global community lies not just in the innovations it creates but in the sense of belonging and purpose it fosters. By coming together, sharing our stories, and working towards a common goal, we are not just building a community; we are shaping the future.

Social Entrepreneurship and Innovation

In the kaleidoscope of modern work cultures, social entrepreneurship, and innovation do not just act as buzzwords but form the bedrock for a transformative movement. This spectrum of activity is where passion meets pragmatism, where visionaries dare to dream of a future that merges prosperity with purpose. The DIY Factory and New Work movements have taken a bold step into this realm, embodying the spirit of innovation and social entrepreneurship at their core.

The essence of social entrepreneurship lies in identifying societal problems and addressing them with innovative solutions. It's about creating businesses that serve a greater good, bridging the gap between profit and social impact. Within the DIY Factory and New Work movements, this translates into crafting workspaces and employment structures that not only foster creativity and productivity but also contribute positively to the community and environment.

Innovation, in this context, isn't confined to technological advancements alone. It also encompasses innovative ways of thinking about work itself, how it's organized, and how it impacts the lives of those it touches. The DIY Factory model, with its emphasis on autonomy, creativity, and collaboration, serves as a fertile ground for such innovation to flourish.

The fusion of social entrepreneurship and innovation within the New Work movement paves the way for a more inclusive and equitable work culture. It challenges the traditional paradigms of corporate hierarchies and rigid job roles, advocating instead for a model where individuals have the freedom to explore and express their unique talents and passions.

This shift towards a more human-centric approach to work encourages not only personal fulfillment but also drives societal progress. When people are empowered to pursue work that aligns with their values and aspirations, they're more likely to produce work that benefits the broader community. This alignment of individual and collective goals is a hallmark of social entrepreneurship.

The DIY Factory, as a physical and conceptual space, embodies these principles. It functions as a hub for innovation, where resources and knowledge are shared openly, fostering a culture of continuous learning and improvement. Here, individuals come together to experiment, prototype, and implement solutions to real-world problems.

Jordan Grey

Implementing these innovative practices in professional environments requires a shift in mindset from both leaders and teams. It calls for leaders who are not just visionaries but also facilitators and supporters of their team's intrinsic motivations and creative processes. It's about creating an environment where failure is not feared but seen as a necessary step towards innovation.

Furthermore, the New Work movement emphasizes the importance of flexibility and autonomy. By providing individuals with the control over when, how, and what they work on, businesses can unlock unprecedented levels of creativity and productivity. This autonomy also supports a better work-life balance, contributing to overall well-being and job satisfaction.

However, adopting these practices is not without its challenges. Traditional industries and organizations might find it difficult to break free from established norms and embrace these new paradigms. It requires a concerted effort from all stakeholders to understand the potential benefits and work collaboratively towards implementing them.

One of the key strategies in overcoming these barriers is through education and skill development. By rethinking education to focus more on creativity, innovation, and entrepreneurial skills, we can prepare future generations for the realities of the New Work movement. It's about encouraging a mindset of lifelong learning and adaptability.

Another important aspect is the role of technology. In the realm of the DIY Factory and New Work, technology serves not just as a tool for efficiency but as an enabler of creativity and collaboration. From digital fabrication tools to virtual collaboration platforms, technology makes it possible to bring innovative ideas to life and connect with like-minded individuals around the globe.

The economic impact of these movements cannot be understated. By shifting the focus from consumerism to production, from centralized manufacturing to decentralized, community-focused production hubs, we can create more sustainable economic models. These models not only support local economies but also contribute to a reduction in environmental impact.

Legal and ethical considerations also play a crucial role in the successful implementation of these models. From navigating the complexities of the gig economy to ensuring intellectual property rights and open-source collaboration, these issues require careful consideration and innovative solutions.

At the heart of it all, the DIY Factory and New Work as social movements represent a beacon of hope for the future of work. They offer a vision of a world where work is not just a means to an end but a source of fulfillment and positive societal impact. It's a call to action for entrepreneurs, professionals, and enthusiasts to come together and work towards making this vision a reality.

In conclusion, the intersection of social entrepreneurship and innovation within the DIY Factory and New Work movements is not just a trend but a testament to the indomitable human spirit. It's a journey of transformation, of challenging the status quo, and daring to imagine a better future for work and society. And it's a journey that we all have a role to play in.

Chapter 20:
Overcoming Barriers to Adoption

Embracing the synergy of DIY factories and New Work concepts represents a seismic shift from traditional paradigms, one that naturally encounters skepticism and resistance. This chapter delves into the heart of these challenges, spotlighting the mindset shifts and strategic actions necessary to make innovation a reality in today's work environments. Overcoming barriers is less about dismantling objections point-by-point, and more about inspiring a vision of what could be. It's in the storytelling of turnaround success stories, the methodical approach to incremental change paired with a bold overarching vision, that the path forward emerges. For forward-thinking professionals, entrepreneurs, and enthusiasts, the journey will involve converting skeptics through demonstrating value, navigating through the friction points of adoption with empathy and understanding, and ultimately cultivating an environment where innovation thrives. This chapter provides the map to navigate this terrain, outlining strategies to shift perspectives, foster buy-in at all organizational levels, and transform challenges into stepping stones towards a more innovative, productive, and fulfilling future.

Addressing Skepticism and Resistance

In the journey toward embracing the principles of DIY factories and New Work, leaders and innovators will inevitably face skepticism and resistance. This pushback is not just an obstacle but a vital opportunity to engage in meaningful dialogue and demonstrate the tangible benefits of these models. It's crucial to understand that resistance often stems from a fear of the unknown or a misalignment between one's values and the perceived impact of change. Overcoming this begins with clear, empathetic communication and the sharing of success stories where similar challenges were met with transformative results. By creating an environment that values curiosity over conformity, one can gradually shift perceptions. Encourage participation in small-scale projects or pilot programs to offer firsthand experience with the new models. This allows skeptics to see, feel, and understand the potential benefits in a low-risk setting. Patience and persistence in these efforts can convert even the most steadfast doubters into champions of innovation, forging a path toward a more adaptable, creative, and fulfilling professional environment.

Case Study: Turnaround Success Stories In the realm of transformative business models, the stories of companies that have pivoted from conventional methods to adopt the principles of DIY factories and New Work are not just inspiring—they're a beacon for the future of work. Through adversity, these organizations have embraced change, redefined their cultures, and emerged more resilient and innovative. Let's delve into these narratives, drawing lessons and motivation for those poised to make their own waves in their industries.

The first emblematic case comes from a technology firm that was on the brink of obsolescence. Hampered by traditional hierarchies and slow innovation cycles, the company's relevance and market share were dwindling. The turning point was its audacious decision to restructure its entire operation based on New Work principles. By flattening its hierarchy, the company fostered open communication and rapid decision-making. Incorporating DIY factories within its R&D departments enabled hands-on innovation, directly involving engineers and designers in the prototyping process. This shift not only revitalized their product line but also reinvigorated the workforce's morale and creativity.

Another remarkable story unfolds within the manufacturing sector. A traditional manufacturing entity facing global competition and rising production costs embarked on a transformative journey. Leveraging the concept of DIY factories, it decentralized its production processes. This enabled local teams to tailor products more closely to regional market demands and substantially cut down logistics costs. Simultaneously, by adopting New Work practices, employees at all levels were empowered to contribute ideas and improvements, leading to a surge in efficiency and job satisfaction. The result was a remarkable turnaround in productivity and profitability, proving the viability of New Work principles in a sector often viewed as rigid.

In the realm of retail, a once-dominant player found itself struggling against the rise of e-commerce giants. The solution came through an embracing of community and innovation. By transforming some of its spaces into communal DIY factories and workshops, they created a new value proposition for customers: not just to buy, but to create. Coupled with an internal overhaul based on New Work values, where staff took on more creative and customer-engagement roles, the company revitalized its brand. It became a hub for makers, artists, and entrepreneurs, showcasing the power of community and experiential retail.

The publishing industry, too, provides a telling example. A mid-sized publisher, grappling with declining sales, turned its fortunes around by becoming a platform for collaborative writing and publishing. Internally, it adopted New Work principles to encourage autonomy and creativity among its employees. Externally, it offered tools and workshops for writers to co-create content, leveraging crowdsourcing and community involvement. By doing so, it not only diversified its revenue streams but also positioned itself as an innovative leader in a changing industry.

In all these cases, critical to success was leadership's willingness to embrace radical change and its commitment to the principles of New Work and DIY ethos. Leaders acted as facilitators rather than traditional bosses, fostering environments where innovation could thrive. They showed that, with vision and courage, turning around a struggling company is more than possible—it's an opportunity to lead the charge towards a future where work is more flexible, creative, and fulfilling.

These turnaround success stories echo a common theme: the transformative power of embracing New Work practices and the DIY factory model. They demonstrate not only survival but also thriving through innovation, adaptability, and a profound reimagining of what work can be. As we look to the future, these narratives offer not just a roadmap but also hope and inspiration for organizations and individuals alike, ready to rethink the status quo and carve out their path in the New Work era.

The implications for forward-thinking professionals, entrepreneurs, and enthusiasts are clear. There's a tangible shift towards environments that value creativity, adaptability, and meaningful work. Harnessing the principles of New Work and the DIY ethos isn't merely a strategy for organizational revival. It's a template for building resilient, innovative, and humane workplaces poised for the challenges and opportunities of the future.

As exciting as these stories are, they're merely the heralds of what's possible when we dare to reimagine the fabric of work itself. The journey of transformation is fraught with challenges, but as shown, the rewards are substantial. It's about crafting not just a more successful business but also a more fulfilling work culture that resonates deeply with the aspirations of today's workforce.

In conclusion, the stories of these turnaround successes are not outliers or anomalies. They are harbingers of a shift in the global work paradigm. They spotlight the immense potential for organizations willing to break free from conventional molds and courageously adopt New Work and DIY factory principles. For those on the brink of transformation, these stories serve as both a guide and inspiration, illuminating the path towards a future where work is synonymous with innovation, flexibility, and personal fulfillment.

Charting a Path Forward: Incremental Change and Big Vision

As we peek over the horizon of traditional workspace models, the blueprint for the future of work becomes evident: a landscape marked by DIY factories and New Work principles. The crux of navigating this transformation lies not merely in the grandeur of vision but in the meticulous orchestration of incremental changes that cumulate into a seismic shift in work culture.

The journey towards adopting innovative practices requires a blend of both patience and audacity. It's about laying one brick at a time while keeping your eyes fixed on the edifice you aim to build. The barriers to adoption aren't insurmountable; they're merely stepping stones, awaiting the right strategies to convert them into milestones.

Understanding that resistance often stems from skepticism, the first step is to cultivate an environment that welcomes questions and critical thinking. It's about fostering a culture where skepticism is not seen as a deterrent but as a springboard for dialogue and deeper understanding. Creating platforms for sharing success stories can significantly mitigate resistance, showcasing tangible examples of transformations achieved through New Work and DIY factory principles.

Incremental change begins with acknowledging the current state and then taking deliberate, albeit small, steps towards the envisioned future. This might mean initiating pilot projects or prototypes within traditional industries to illustrate the benefits of more flexible, creative, and innovative work models. It's these minor adjustments and experiments that pave the way for broader acceptance and adoption.

Yet, to truly realize the vision of DIY factories and New Work concepts, a larger, more encompassing perspective is essential. This vision encompasses not only the transformation of individual workspaces but also the redefinition of economic, environmental, and social paradigms. It's about envisioning a future where work and creativity intersect seamlessly, generating not only economic value but also contributing to social and environmental well-being.

Leadership plays a pivotal role in bridging the gap between incremental change and big vision. Leaders who embody the principles of New Work and who are committed to the ethos of the DIY factory model become catalysts for change. By leading by example, they inspire innovation, creativity, and a willingness to experiment and take risks.

Empowering teams is another critical facet of charting a path forward. It's about decentralizing decision-making and giving individuals and teams the autonomy to explore, create, and innovate. This empowerment fosters a sense of ownership and accountability, crucial for nurturing a culture that embraces change and innovation.

Education and continuous learning emerge as the backbone of successful transition. As workplaces evolve, so do the skills required to thrive within them. Thus, investing in education and skill development is non-negotiable. This investment isn't limited to technical skills; it extends to nurturing creativity, critical thinking, and the ability to collaborate across traditional boundaries.

Technology, too, plays a defining role in this journey. Embracing technologies that enhance efficiency, foster creativity, and facilitate collaboration is fundamental. Yet, it's equally important to remain vigilant about the potential pitfalls of over-reliance on technology. Balancing technological advancement with human insight and creativity is key.

The economic models that underpin DIY factories and New Work principles demand careful consideration. Shifting from consumerist models to ones that emphasize production, sustainability, and shared value creation is an incremental process. It involves challenging existing economic paradigms and experimenting with models that prioritize long-term value over short-term gains.

Community engagement and building partnerships stand out as indispensable elements of this transformation. The journey towards a new work culture is not a solitary endeavor; it requires collaboration and co-creation. Engaging with local communities, fostering partnerships across sectors, and building networks of like-minded practitioners amplify the impact of incremental changes and contribute to realizing the broader vision.

Moreover, the path forward requires a reevaluation of success metrics. Moving beyond traditional indicators to metrics that capture impact, engagement, sustainability, and innovation is crucial. These new metrics offer a more holistic view of progress, one that aligns with the principles of New Work and the ethos of DIY factories.

To ensure that this vision doesn't remain a distant dream, advocacy and policy reform play a pivotal role. Advocating for policies that support flexibility, innovation, and the shift towards new economic models is essential. It's about creating an ecosystem that nurtures the growth of DIY factories and the adoption of New Work principles, making them accessible and viable for a broader spectrum of society.

Ultimately, charting a path forward is an exercise in balance. It's about balancing incremental change with a steadfast commitment to a larger vision. It's about balancing scalability with sustainability, technological advancement with human-centricity, and economic efficiency with social and environmental responsibility. This balanced approach ensures that the journey towards innovative work models is both attainable and sustainable.

In conclusion, the transformation towards DIY factories and New Work principles is both an evolutionary and a revolutionary process. It requires the courage to reimagine the future of work, the patience to implement incremental changes, and the wisdom to navigate the complexities of this transformation. The path forward is carved by the collective efforts of leaders, innovators, and communities worldwide, each contributing to a future where work is not only productive and innovative but also deeply fulfilling and meaningful.

Chapter 21:
The Intersection of Art, Culture, and Work

In an era where the boundaries between work and life blur, we dive into the harmonious blend of art, culture, and work that paints the canvas of our professional landscapes. Art is not just a domain for creators outside of the corporate sphere; it's a vital pulse within it, driving innovation and fostering an environment ripe for the cultivation of groundbreaking ideas. At the core of the DIY factories and New Work movements is an understanding that culture is not just a backdrop for work—it's a dynamic framework that shapes it. When artistic endeavors are integrated into the fabric of our working lives, they catalyze creative expression in forms we hadn't imagined, making the mundane magnificent and transforming challenges into opportunities for growth. This cross-pollination of artistic and cultural sensibilities with our work practices encourages a more holistic approach, inviting diversity of thought and promoting a culture of continuous exploration. By embracing the intersection of art, culture, and work, we're not merely adopting new methods; we're redefining the essence of productivity to include vibrancy, meaning, and a profound connection to the human spirit. It's a transformative journey that goes beyond mere aesthetic appeal, anchoring deeply into how we view collaboration, innovation, and the endless possibilities of what we can achieve when we dare to blend the seemingly disparate worlds of art, culture, and work.

Creative Expression as a Workforce Paradigm

In the transformative landscape of modern work, the surge towards integrating creative expression within the workforce heralds a paradigm shift of monumental proportions. At the heart of this movement lies the recognition that creativity isn't just an asset but a fundamental principle that fuels innovation, problem-solving, and collaboration. Empowering employees to channel their artistic sensibilities within the workplace doesn't merely enhance aesthetic value; it cultivates an environment where unique ideas flourish, engagement deepens, and resilience strengthens. This philosophy invites a reimagining of roles where individuals contribute not only through their skillsets but also through their creative insights, thereby fostering a culture that champions diversity of thought and innovation. By embracing creative expression as a core component of the workforce, organizations set the stage for breakthrough achievements, demonstrating that when people feel valued for their full range of abilities, they're not just more productive, they're profoundly more fulfilled and connected to their work. Thus, navigating the intersection of art, culture, and work doesn't just enrich our professional landscapes; it propels them towards a future where work is not only about what we do but also celebrates who we are.

Artistic Endeavors Within DIY Factories Within the innovative hubs of DIY factories, a unique fusion of art and manufacturing breathes new life into what many had seen as separate realms. These spaces aren't just about output and efficiency; they stand as bold declarations that the heart of creation beats not only in the artist's studio but also within the mechanical hum of the factory floor. This chapter delves into the colorful intersection where art meets industry, crafting a narrative that might just inspire a renaissance in the modern workspace.

The essence of a DIY factory is innovation and flexibility, qualities that naturally dovetail with the artistic process. Picture a space where the tools of production – from 3D printers to laser cutters – are not confined to creating prototypes for the market but are also available to those looking to express a creative vision. It's here that you'll find artists and engineers, side by side, pushing the boundaries of what's possible.

Imagine the potential when art is not seen as merely decorative but integral to the product's design and function. In these environments, artists collaborate with product designers to create items that are both functional and aesthetically striking. This collaborative process not only enriches the product but also imbues it with a story, a character that stands out in a crowded marketplace.

Art within the DIY factory also serves as a bridge to the community. Many factories open their doors to local artists, providing them with access to tools and technologies that might be otherwise out of reach. This symbiosis can lead to community-driven projects, workshops, and events that foster a vibrant local art scene, turning the factory into a cultural hub.

This blending of disciplines invites a new kind of worker into the industrial fold: the artist-technologist. These individuals are as comfortable coding or using a CNC machine as they are with a paintbrush or clay. Their presence in the DIY factory not only enriches the creative pool but also exemplifies the versatility that the future workforce will need.

One of the most powerful aspects of incorporating art into the factory setting is the effect on innovation. Art challenges norms, pushes boundaries, and encourages thinking outside the box. When this kind of creative thinking permeates a workspace, the potential for groundbreaking ideas and products skyrockets. It's no longer about incremental improvements but about leaps of imagination that redefine what's possible.

The aesthetics of the workspace itself can greatly benefit from an artistic touch. Gone are the sterile, utilitarian environments of traditional factories. In their place, vibrant, thoughtfully designed spaces that stimulate creativity and foster well-being. These are places where people want to be, blurring the line between work and play.

Projects born in DIY factories that embrace artistry often carry a unique signature. They tell a story not just of utility but of culture, craftsmanship, and human connection. These projects resonate on a deeper level with consumers, standing as testaments to the blending of form and function.

Moreover, the convergence of art and manufacturing opens up new market opportunities. Customization and personalization, attributes highly valued in today's market, are naturally integral to artistic endeavors. This can give DIY factories a competitive edge, offering products that are not only innovative but deeply reflective of individual or community identity.

Environmental sustainability is another area where the artistic endeavors within DIY factories shine. Artists often repurpose materials, bringing attention to recycling and sustainability issues through their work. This mindset encourages factories to think more critically about their materials and processes, driving towards greener, more sustainable production.

Education and skill development benefit greatly from this artistic infusion. Workshops and training sessions led by artists can demystify technology for the uninitiated, breaking down barriers and fostering a more inclusive environment for innovation. These educational opportunities create a pipeline of talent, skilled not just in the technical aspects of production but also in creative problem-solving.

The artistic endeavors within DIY factories also contribute to a richer, more diverse company culture. They promote an environment where experimentation is celebrated, failure is seen as a learning opportunity, and diversity of thought is encouraged. This not only leads to happier employees but also to more resilient and innovative organizations.

The impact of art in the DIY factory extends beyond its walls. As products and projects reach the wider community, they can challenge perceptions about manufacturing, technology, and art. They demonstrate that these arenas are not mutually exclusive but are interconnected and enriched by each other's presence.

In the end, the integration of artistic endeavors within DIY factories represents more than simply a blending of disciplines. It signifies a broader shift in how we perceive work, creativity, and the very definition of innovation. These spaces stand as models for the future, where the collective talent of artists, engineers, and entrepreneurs can come together to create not just products but a more vibrant, thought-provoking world.

The call to action for professionals, entrepreneurs, and enthusiasts is clear: embrace the artistic within the industrial. Explore the unlimited potential that comes from breaking down the walls between disciplines. Through this convergence, we can redefine not just what we make, but how and why we make it, paving the way for a future that is both innovative and inspiring.

The Cultural Impact of New Work Movements

In an era where the lines between art, culture, and work blur more each day, new work movements have emerged as powerful catalysts for change. At the heart of this transformation is a belief that work isn't just a means to an end but a conduit for creative expression and societal impact. This section explores the profound cultural impact of these movements, weaving together the narratives that define our contemporary work landscape.

Historically, work has been compartmentalized, separated from the realms of personal passion and creativity. Yet, the rise of new work movements challenges this notion, arguing for a holistic approach where work is an integral part of one's identity and life's purpose. The DIY ethos and the principles of New Work have not only redefined what it means to work but have also reshaped societal norms and expectations surrounding employment.

The influence of new work movements extends beyond individual fulfillment, seeding changes in community dynamics and global cultures. By valuing creativity, autonomy, and collaboration, these movements foster environments where diverse talents can converge, sparking innovation and inspiring a shared vision for the future. This cultural shift towards a more inclusive and participative work ethos reflects a deeper desire for connection and meaning in our professional lives.

Art and culture have always been mirrors reflecting societal values and priorities. As new work movements gain momentum, they infuse work cultures with artistic sensibilities, challenging the conventional wisdom that art and work are mutually exclusive. In DIY factories and innovative workspaces around the globe, artistry and creativity are not just welcomed but regarded as essential to problem-solving and value creation.

This fusion of art, culture, and work is producing new genres of professional endeavors and reimagining what workplaces look like. Spaces are designed not just for efficiency but as canvases for creative expression, fostering environments where inspiration can flow freely. From office murals to design-thinking workshops, the aesthetic and cultural dimensions of workspaces are being transformed.

The implications of these shifts are profound, reaching into the fabric of communities and economies. As new work practices prioritize sustainability, social responsibility, and ethical engagement, they're setting new standards for how businesses operate and compete. This cultural turn towards more mindful and purpose-driven work is reshaping consumer expectations and influencing global market trends.

Moreover, the rise of these movements has democratized opportunities for creative expression, challenging the traditional barriers to entry in many fields. The accessibility of technologies and collaborative platforms has empowered a new generation of artists, entrepreneurs, and innovators, who bring diverse perspectives and skills to the fore. This inclusivity enriches the cultural landscape, fostering a more vibrant and dynamic society.

Education and skill development are also undergoing a transformation, as traditional pedagogies give way to more experiential and collaborative learning methods. This shift is preparing future generations for the complexities of modern work, where creativity, adaptability, and lifelong learning are key. The impact of this educational evolution will reverberate through culture, as new generations bring a different set of values and expectations to their professional and personal lives.

The global dissemination of new work principles is facilitating cross-cultural exchanges and collaborations, breaking down geographical and cultural barriers. This global network of thinkers, makers, and creators is not just sharing ideas but co-creating solutions to some of the world's most pressing challenges. Such collaboration is a testament to the unifying power of shared cultural and work-related values.

Yet, the cultural impact of new work movements is not without its challenges. As these ideals proliferate, they also face skepticism, resistance, and the inevitable tensions that arise from challenging the status quo. Navigating these obstacles requires resilience, foresight, and a commitment to the underlying principles of these movements. It is through overcoming these challenges that new work movements can truly instill lasting cultural change.

In the face of uncertainty and rapid change, the cultural narrative around work is evolving. New work movements offer a vision of the future where work is not just a source of economic survival but a key driver of cultural enrichment and social progress. This vision, while still unfolding, highlights the potential for work to elevate human dignity, foster community, and contribute to a more equitable and sustainable world.

As we move forward, it's essential to continue exploring and questioning the cultural implications of these changes. How will they shape our identities, our communities, and our global society? The journey of integrating art, culture, and work is just beginning, and its trajectory will depend as much on our collective imagination as on our willingness to reimagine the possibilities of what work can be.

The cultural impact of new work movements is a testament to the power of human creativity and vision. By embracing these principles, we can not only transform the landscape of work but also enrich our cultural heritage, leaving a legacy of innovation and ingenuity for future generations. The promise of these movements is vast, and their full realization will require a continued commitment to pushing boundaries, challenging norms, and envisioning a future where work and culture are inextricably linked.

In conclusion, the cultural impact of new work movements is reshaping our world, offering new lenses through which to view the role of work in society. As we navigate this evolving landscape, let us embrace the opportunities for creativity, innovation, and transformation that these movements present. By doing so, we not only enrich our own lives but also contribute to a richer, more vibrant cultural fabric for all.

Chapter 22:
Financial Models for Sustainability and Growth

In the journey towards building sustainable and scalable DIY factories and New Work environments, navigating the financial landscape is paramount. This chapter embarks on unraveling the various financial models that not only ensure the survival of these innovative ventures but also pave the way for their exponential growth. We delve into the mechanisms of funding and monetizing innovation, exploring the synergies between crowdfunding, venture capital (VC), and bootstrap models. Each model comes with its unique set of advantages and challenges, but the ultimate goal remains the same: to fuel sustainable development and foster an ecosystem of economic resilience. By strategically leveraging these financial models, entrepreneurs and innovators can unlock new avenues for diversification and stability, ensuring that their ventures stand the test of time. As we dissect these models, it becomes evident that the path to financial sustainability and growth in the DIY factory and New Work domains is not linear. It requires a blend of creativity, adaptability, and strategic foresight. This chapter serves as a guide, empowering readers with the knowledge and tools to navigate the financial intricacies of their innovative endeavors, ensuring that they not only survive but thrive in the ever-evolving landscape of modern work culture and innovation.

Jordan Grey

Funding and Monetizing Innovation

In an era where traditional financial models are increasingly coming under scrutiny for their limitations in fostering genuine innovation, pioneering entrepreneurs and forward-thinkers are exploring alternative avenues to fund and monetize their inventive endeavors. The key to unlocking sustainable growth and ensuring the longevity of your innovations lies not just in the novelty of your ideas but in strategically leveraging a blend of crowdfunding, venture capital, and bootstrap models to fuel your projects. This approach not only maximizes resource efficiency but also aligns financial investment with the ethos of community engagement and participatory development that is central to the DIY factory and New Work movements. Imagine mobilizing a passionate community that not only supports your vision but also contributes to it financially, creating a robust economic ecosystem where innovation thrives. This section delves into how you can transcend traditional funding models, inspiring you to harness the collective power of your community and stakeholders to realize your visionary projects. In doing so, you pave the way for a future where financial sustainability and growth are not just aspirations but tangible realities, achieved through the creative reimagining of funding and monetization strategies.

Crowdfunding, VC, and Bootstrap Models As we delve into the financial models vital for the sustainability and growth of DIY factories and new work initiatives, it's essential to understand the distinct pathways through which these innovative projects can secure the necessary capital. Crowdfunding, venture capital (VC), and bootstrapping represent three pivotal financial avenues, each with its unique set of advantages, challenges, and suitability depending on the nature of the project at hand.

Firstly, let's explore crowdfunding. This method has democratized the funding process, enabling entrepreneurs to raise small amounts of

176

money from a large number of people, typically via the internet. Crowdfunding platforms such as Kickstarter and Indiegogo have become indispensable in validating ideas, engaging with potential customers, and generating the initial capital needed to kickstart innovative projects. For DIY factories and new work concepts, crowdfunding not only provides the necessary funds but also builds a community of supporters who believe in the vision and are eager to see it come to fruition.

However, crowdfunding is not without its challenges. Successful campaigns require a compelling story, an attractive reward system for backers, and rigorous marketing efforts. The pressure to deliver on promises within a specified timeline can be daunting, especially for projects that encounter unforeseen complications in the development phase.

On the other hand, venture capital presents a more traditional route to securing significant financial investments. VCs are on the lookout for high-growth potential startups, where they can inject substantial sums of money in exchange for equity. For innovative ventures that aim to scale rapidly, VC funding can provide not just the necessary capital but also strategic guidance, industry connections, and access to a broader network of resources.

Yet, VC funding is not a one-size-fits-all solution. The process of securing venture capital is highly competitive, with a focus on high returns on investment. This model may not be the best fit for projects prioritizing social impact over profitability, or those wishing to maintain greater control over their company's direction and values.

Bootstrapping stands out as a self-funding strategy, where entrepreneurs rely on their savings, or the initial cash flow generated by the business, to support its growth. This model champions self-reliance and minimizes debt, allowing founders to retain full ownership and control over the direction of their venture.

Bootstrapping is especially appealing to DIY factory and new work proponents who value autonomy and gradual, organic growth.

Despite its advantages, bootstrapping can impose limitations on the speed of growth and the scale at which the business can expand. The constant juggle between investing in the project and maintaining operational costs can be challenging, often requiring a frugal and inventive approach to resource management.

In considering these financial models, entrepreneurs must assess their project's goals, scale, and values. Crowdfunding can catalyze community-building and engagement, helping to validate the project in its early stages. Venture capital is suited for startups with a clear path to rapid growth and scalability, offering not just funding but also mentorship and access to a vast network. Bootstrapping appeals to those prioritizing independence and gradual growth, allowing founders to steer their ventures without external pressures.

It's important to note that these models are not mutually exclusive. Many successful ventures have navigated their growth by leveraging a combination of these funding strategies at different stages of their development. For instance, a project may begin with crowdfunding to validate its concept and then seek VC investment for scaling operations, all while maintaining a bootstrap mentality to manage resources efficiently.

For DIY factories and new work initiatives, understanding these financial pathways is crucial. It enables founders to strategically navigate the complexities of funding, ensuring their innovative projects are not just born but also nurtured to reach their full potential. Navigating the financial landscape requires a balance between powerful storytelling, strategic planning, and prudent financial management.

As we move forward, embracing the DIY ethos and new work paradigms, let's recognize the importance of financial sustainability. It's about making calculated choices that align with our values, vision, and the communities we aim to serve. The journey from idea to impact is fraught with challenges, but with the right financial model, passion-driven projects can thrive, reshaping our world one innovation at a time.

In the end, whether you're rallying a community through a crowdfunding campaign, pitching to discerning venture capitalists, or meticulously bootstrapping your way to sustainable growth, the objective remains the same. It's about bringing innovative ideas to life, fostering a culture of creativity and resilience, and empowering a new wave of entrepreneurs and thinkers to challenge the status quo. By understanding and thoughtfully choosing the right financial path, we lay the foundation upon which groundbreaking projects can flourish, marking the beginning of a new era in work and manufacturing.

Economic Resilience and Diversification

In today's rapidly changing economic landscape, resilience and diversification aren't just buzzwords—they are essential pillars for sustainability and growth. For ventures embracing DIY factory principles and new work innovations, these concepts are even more critical. Their unconventional approaches to value creation require a robust financial backbone that can weather economic downturns and adapt to shifting market demands.

At the heart of economic resilience lies the ability to bounce back from setbacks. This trait is all the more important for modern entrepreneurs and innovators who often operate in environments of uncertainty and risk. The essence of resilience in this context is not just about survival but thriving amidst challenges. It's finding opportunities where others see obstacles, leveraging the unique strengths of the DIY and New Work movements to create value in unexpected ways.

Diversification, on the other hand, speaks to the spreading of risk and opportunity across multiple revenue streams. For those steeped in the DIY factory ethos and new work practices, this could mean exploring various business models, from product sales to consulting services, workshops, and digital content creation. The digital age has democratized access to global markets, making diversification more accessible but also more necessary than ever.

One core strategy for achieving economic resilience and diversification is innovation in product and service offerings. The dynamic nature of DIY factories, combined with the fluidity of new work principles, offers a fertile ground for such innovation. By constantly iterating and experimenting, businesses can stay ahead of market trends and customer needs, thereby securing their place in an ever-evolving economy.

Another strategy is leveraging community and collaboration. The ethos of the DIY and New Work movements is deeply rooted in collective action and shared knowledge. By building strong networks, entrepreneurs can tap into a wider pool of skills, resources, and opportunities. Collaborative projects can not only open new revenue streams but also enhance the resilience of the community as a whole, creating a buffer against economic shocks.

Financial prudence is also paramount. In the pursuit of growth and innovation, it's easy to overlook the fundamentals of sound financial management. Yet, it's often the ventures with stringent budget controls, prudent investment strategies, and robust financial planning that can navigate through tough times. For the DIY factory and New Work enthusiast, this might mean bootstrapping over seeking external funding to retain control and flexibility or exploring crowdfunding as a way to validate products and engage potential customers.

Moreover, embracing technology and digital transformation can play a significant role in bolstering economic resilience. Automation, digital marketing, e-commerce, and remote work technologies not only streamline operations but also open up new avenues for growth and diversification. They allow businesses to scale efficiently, reach global markets, and adapt to digital trends, all of which contribute to a more resilient economic foundation.

Yet, it's important to remember that resilience and diversification are not just about strategic maneuvers and operational tactics. They are also about cultivating a mindset of adaptability, curiosity, and relentless learning. In an age of rapid change, those who are willing to continuously evolve, question the status quo, and embrace uncertainty are the ones who thrive.

Sustainability should also be at the core of resilience and diversification efforts. The intersection of DIY factories and New Work is not just an economic or technological revolution but a cultural and social one as well. By prioritizing environmentally sustainable practices and socially responsible business models, ventures can build long-term viability into their DNA. This not only garners support from like-minded customers and partners but also aligns with the global shift towards sustainability.

Furthermore, the pursuit of economic resilience and diversification should be inclusive, ensuring that the benefits of growth and innovation are shared across society. This inclusivity can extend the reach and impact of the DIY and New Work movements, creating opportunities for communities that have traditionally been marginalized in the industrial and digital economies.

Jordan Grey

In practice, building economic resilience and diversification can take many forms. For some, it may involve pivoting to new markets or products in response to changing consumer demands. For others, it might mean forming strategic partnerships to enter new sectors or leveraging digital platforms to reach a global audience. The key is to remain open to change, responsive to opportunities, and grounded in a deep understanding of one's values and capabilities.

As we forge ahead, the landscapes of work and manufacturing will continue to evolve, influenced by technological advancements, global economic shifts, and societal changes. The principles of resilience and diversification will remain central to navigating these changes successfully. They will enable entrepreneurs, innovators, and forward-thinkers to not only withstand the winds of change but to harness them, crafting a future of work that is sustainable, inclusive, and brimming with opportunity.

In conclusion, economic resilience and diversification are not merely strategies but essential philosophies for those at the forefront of the DIY factory and New Work movements. By embedding these principles into the fabric of their ventures, innovators can pave the way for a new era of economic stability, creativity, and growth. As we embrace these concepts, we step into a future where work is not just a means to an end but a continuous journey of exploration, challenge, and fulfillment.

In this journey, the role of the individual entrepreneur, the community, and technology intertwine to create a resilient and diversified economic model that not only withstands challenges but also thrives on them. This is the future we are crafting—a future built on the foundations of innovation, collaboration, and a profound commitment to sustainable growth and development.

Chapter 23:
Measuring Success in the New Work Economy

In the fast-evolving landscape of the new work economy, the traditional metrics of success—revenue, profit margins, and market share—remain important but no longer capture the full picture. As we delve into the nuances of measuring success in this innovative ecosystem, we shift our focus towards more holistic and forward-thinking indicators. Success in this realm is increasingly defined by the value created for a wider community, the engagement and well-being of teams, and the sustainable impact on the environment. It's not just about what is achieved, but how it's achieved and who benefits. The blend of DIY factory principles with new work concepts pushes us to re-evaluate our metrics, emphasizing continuous improvement, adaptability, and the fostering of a creative and autonomous workforce. This chapter explores how embracing unconventional metrics such as impact, engagement, and legacy can not only drive businesses and individuals towards more meaningful achievements but also foster environments where innovation thrives. We argue that assessing success through these lenses encourages a culture of learning, resilience, and community-oriented growth. As we deconstruct these new metrics, it's clear that the journey towards innovation and sustainability is ongoing, emphasizing the importance of openness to change and the relentless pursuit of excellence in both personal and professional domains.

Beyond Traditional KPIs: New Metrics for a New Era

In the ever-evolving landscape of the new work economy, traditional key performance indicators (KPIs) no longer suffice to capture the full spectrum of innovation, collaboration, and impact that modern enterprises strive for. It's high time we pivot our metrics towards those that encapsulate the essence of value creation in this new era. Embracing a new set of KPIs requires a deep dive into measures beyond mere financials, focusing on impact, engagement, and legacy—indicators that truly reflect the entrepreneurial spirit and collaborative ethos inherent in the DIY and New Work movements. These metrics highlight the significance of building a sustainable community, fostering continuous learning, and championing environmental responsibility. They prompt us to question, "How are our endeavors not just profitable, but also beneficial to society and nurturing for our collective future?" By incorporating these broader, more inclusive metrics into our success paradigms, we chart a course towards a work environment that prioritizes creativity, innovation, and meaningful contribution, thereby inspiring professionals and entrepreneurs to not only dream bigger but to act in ways that leave a lasting, positive impact on the world.

Impact, Engagement, and Legacy The transformative power of combining the DIY ethos with New Work principles lies not just in the novel ideas and innovative solutions it spawns but also in the profound impact it has on individuals, organizations, and communities at large. By embracing this synergy, we're not merely adopting a set of practices; we're embarking on a journey that reshapes our understanding of work, creativity, and collaboration. The effects of this transformation extend far beyond the confines of any single DIY factory or organization embracing New Work principles. They ripple out to influence the broader economic, social, and cultural context in which we operate.

When individuals are empowered to take ownership of their creative processes and work environments, the level of engagement and intrinsic motivation skyrockets. This heightened engagement is not a fleeting phenomenon. Instead, it becomes the fuel that drives innovation and continuous improvement. Organizations that recognize and foster this reality find themselves at the forefront of their industries, not because they've implemented the latest technologies, but because they've tapped into the most renewable energy source available: human creativity and passion.

Moreover, the legacy of integrating DIY and New Work principles into our professional lives extends well beyond the immediate outcomes of enhanced innovation and productivity. It challenges and ultimately changes the paradigms of leadership, teamwork, and growth. By flattening hierarchies, encouraging open communication, and fostering a culture of mutual respect and continuous learning, we're cultivating environments where the future leaders of innovation are nurtured. These future leaders will be characterized not by their ability to assert control over others but by their capacity to inspire, engage, and collaborate.

However, the impact of this cultural shift isn't confined to the corporate or organizational level. It spills over into the broader community, influencing how we conceptualize the relationship between work and life, between individual achievement and community development. Communities surrounding vibrant DIY factories and organizations practicing New Work become hubs of creativity and innovation, attracting like-minded individuals and companies. This, in turn, creates a virtuous cycle that propels economic development and positions these communities as beacons of progress and resilience in an ever-changing global landscape.

In terms of legacy, one of the most significant contributions of the DIY factory and New Work movement is the redefinition of success.

Success in this new paradigm isn't measured solely by profit margins or market share but by the impact an organization has on the world around it. This includes its environmental footprint, its contribution to the community's well-being, and the ways in which it enriches the lives of its employees. This broader definition of success inspires a new generation of entrepreneurs and leaders who prioritize sustainability, social responsibility, and genuine innovation over short-term gains.

The engagement fostered by this approach also redefines the nature of competition. In the traditional model, companies vie against each other in a zero-sum game where the success of one necessitates the failure of another. Within the ecosystem of DIY factories and New Work environments, however, competition takes on a more collaborative and constructive form. Here, organizations view each other not merely as rivals but as potential partners in innovation. By sharing ideas, technologies, and even failures, they collectively push the boundaries of what's possible, creating a richer, more diverse landscape of solutions and opportunities.

This collaborative approach extends beyond the immediate industrial or technological domain. It influences the broader culture, encouraging a shift away from consumerism toward a more sustainable, production-oriented society. As more individuals gain the skills, confidence, and inspiration to create rather than consume, we begin to see a reduction in waste, a revitalization of local economies, and a greater sense of community cohesion.

The legacy of the DIY and New Work movement is, therefore, one of empowerment. By challenging traditional notions of work, creativity, and collaboration, it empowers individuals to become active creators of their destinies. It empowers organizations to become catalysts for positive change, not just in their industries but in society at large. And it empowers communities to transform into vibrant, sustainable ecosystems where innovation and well-being go hand in hand.

As we look to the future, the full impact of this transformation remains to be seen. What's clear, however, is that the principles of DIY and New Work offer a compelling vision of what the future of work could look like. A future where work is not just a means to an end but a fulfilling, integral part of our lives. A future where organizations don't just create products or services but cultivate environments where creativity, innovation, and well-being flourish.

This vision of the future isn't just aspirational. It's achievable. Across the globe, we're already witnessing the emergence of organizations and communities that embody these principles. Each success story serves as a proof of concept, demonstrating the viability and value of this transformative approach to work.

Embracing the principles of DIY and New Work doesn't just impact our professional lives; it fundamentally alters our relationship with work, creativity, and community. It offers a pathway toward a more sustainable, equitable, and innovative future. As we stand at the threshold of this new era, the choice is ours. We can continue down the well-trodden path of traditional work paradigms, or we can embrace the challenge of reimagining what work could be. For those ready to make that leap, the rewards are immense—not just for ourselves but for generations to come.

In sum, the legacy of intertwining DIY and New Work paradigms is one of profound transformation. By fostering environments where innovation is the norm, where failure is embraced as a stepping stone to success, and where every individual is empowered to contribute their unique talents and perspectives, we're not just changing the way we work. We're shaping a future that values creativity, community, and sustainability above all else. As this movement continues to evolve, its impact and legacy will undoubtedly grow, inspiring and influencing countless individuals and communities around the world.

Ultimately, the true measure of the success of the DIY factory and New Work movement will be the world it helps create. A world where work enriches our lives and our communities, where innovation is driven not by competition but by collaboration, and where the legacy we leave for future generations is a planet that is healthier, societies that are more equitable, and economies that are more vibrant and inclusive. This is the future we're working towards, one innovative project, one empowered individual, and one transformed organization at a time.

Continuous Improvement and Innovation Metrics

In the evolving landscape of work, particularly within the realms of DIY factories and new work environments, success cannot merely be quantified through traditional metrics such as productivity, revenue, and efficiency. While these conventional Key Performance Indicators (KPIs) offer valuable insights, they fall short of encapsulating the full spectrum of innovation and continuous improvement that defines the new work economy. To truly measure success in these progressive settings, we must delve into metrics that resonate with the ethos of continuous innovation and improvement.

Continuous improvement and innovation metrics serve as a beacon, guiding organizations towards fostering cultures of perpetual learning and adaptation. These metrics don't just quantify the tangible; they evaluate the intangible facets of work that are often overlooked but are crucial for sustainable growth and development. Attributes such as employee engagement in creative processes, the rate of innovation adoption, and the breadth of skills acquisition are pivotal indicators of a thriving work environment in the context of new work and DIY factories.

One critical metric is the innovation adoption rate, which measures how swiftly and effectively new ideas or technologies are embraced and implemented within the organization. This metric sheds light on the organization's agility and its capacity to stay relevant in a rapidly changing market landscape. A high adoption rate indicates a responsive and adaptable organizational culture, key traits in the fast-paced world of new work.

Another significant metric is the diversity of skills acquisition. In the pursuit of innovation, the amalgamation of diverse skills and perspectives cannot be overstated. By measuring the range and depth of skills employees acquire over time, organizations can gauge their internal capacity for creativity and problem-solving. This metric is indicative of a culture that values learning and cross-functional collaboration, pillars of the DIY factory and new work philosophy.

Employee engagement in creative processes is also a crucial metric. It assesses the level of active participation and enthusiasm employees have towards involvement in creative endeavors and innovation projects. High engagement levels are synonymous with a vibrant, dynamic work culture that encourages exploration and experimentation. This metric is fundamental in understanding whether an organization's environment is conducive to creative thought and action.

The rate of iterative improvements within projects or processes represents another vital metric. This measures the frequency and effectiveness of incremental changes made in pursuit of excellence. It reflects an organization's commitment to refining and enhancing its outputs, embodying the very essence of continuous improvement. Tracking this metric can highlight a team's agility and resilience in face of challenges, key components in the fabric of new work ecosystems.

Furthermore, the volume and quality of ideas generated within a given timeframe offer insights into the innovative capacity of the organization. This metric is not just about quantity; it is about the potential impact and feasibility of these ideas, indicating the organization's ability to think outside the box and push boundaries.

Customer or end-user feedback on innovative solutions is another essential metric. In the end, the effectiveness of innovation is gauged by the value it creates for its intended beneficiaries. Engaging with feedback mechanisms allows organizations to pivot and adapt, ensuring their innovations meet real-world needs and preferences. This metric aligns closely with the DIY factory ethos of creating meaningful, user-centric solutions.

Employee satisfaction and retention rates in the context of creative work environments also serve as leading indicators. They provide a glimpse into how well the organization supports its employees' creative aspirations and professional development. A high retention rate in a highly innovative environment suggests a strong alignment between the organization's values and the personal aspirations of its workforce.

Moreover, measuring the impact of innovations on society or the environment offers a broader perspective on success. It transcends traditional business metrics to encompass social and environmental responsibility, echoing the new work and DIY factory movements' emphasis on generating positive change.

Implementing these metrics requires a thoughtful approach. Organizations should tailor them to align with their unique goals, culture, and context. This might involve developing new tools or methodologies for data collection and analysis, as well as fostering a mindset that values diverse forms of success.

To effectively track these metrics, organizations must cultivate an environment that encourages transparency, feedback, and critical reflection. This involves not just collecting data, but engaging with it in meaningful ways that drive action and improvement. The goal is to create a feedback loop where insights derived from these metrics inform strategic decisions and initiatives.

In conclusion, measuring success in the new work economy demands a nuanced approach that recognizes the multi-dimensional nature of innovation and continuous improvement. By embracing a broader set of metrics, organizations can paint a more complete picture of their progress, one that mirrors the dynamic, fluid nature of today's work landscape. This shift not only aligns with the principles of DIY factories and new work environments but also sets the stage for sustained growth and impact in the ever-evolving world of work.

Ultimately, the journey towards embracing these metrics is iterative and evolving. It's about continuously refining our understanding of what success looks like in a world where change is the only constant. By integrating these innovative metrics into our evaluation frameworks, we position ourselves to not just navigate but thrive in the new work economy, fostering environments that cherish learning, adaptation, and above all, improvement.

Chapter 24:
Building Your Own DIY Factory and
New Work Ecosystem

In this pivotal chapter, we dive into the transformative journey of building your own DIY Factory and crafting a new work ecosystem that not only empowers but also revolutionizes the traditional notions of productivity and creativity in the modern landscape. By embarking on this venture, you're not merely setting up a physical space for innovation; you're cultivating a mindset and a culture that champions autonomy, collaboration, and continuous learning. Imagine a workspace where every individual has the agency to experiment, where failures are not setbacks but stepping-stones, and where the hierarchy is so flat it practically blends into the collaborative floor. This isn't just an idealistic dream. It's a tangible reality for those willing to challenge the status quo and lead the charge towards a future where work is not only about earning a living but about bringing out the best in everyone involved. Through leveraging technology, embracing open communication, and fostering a community of like-minded innovators, you can turn the keys to unlock a new realm of possibilities. The creation of a DIY Factory and a new work ecosystem is a bold move towards not just personal or organizational growth but towards contributing to a global movement reshaping what it means to work, create, and innovate. Let this chapter serve as your blueprint, your inspiration, and your call to action to construct a foundation that will stand the test of time and change the face of work as we know it.

A Step-by-Step Guide to Launching Your Initiative

In embarking on the monumental journey of establishing your DIY Factory and nurturing a new work ecosystem, it's crucial we lay down a concrete, step-wise path that not only guides but also galvanizes you into action. First and foremost, crystallize your vision by defining what you aim to achieve through your initiative. This clarity will act as your North Star, guiding every decision you make. Next, conduct a thorough market and resources analysis to ensure your factory can thrive and sustain. With your groundwork laid, focus on building a robust blueprint including the physical layout, the technological backbone, and, importantly, the human element – your team. Select individuals not just for skill but for their shared passion for innovation and a new work culture. Following the team assembly, initiate a rigorous phase of planning and execution where your ideas and prototypes take shape. Throughout this process, maintain a laser focus on fostering a culture of collaboration, creativity, and continuous learning. This foundation is not just for the physical aspects of your factory but also the intangible ethos that will define your workspace. Remember, launching your initiative is not the culmination but the beginning of an iterative process of refinement and growth. As you proceed, leverage feedback loops, adapt, and evolve. Your journey from vision to reality is a testament to the power of innovation and resilience, serving as a beacon for future pioneers in the realms of DIY factories and new work ecosystems.

From Vision to Reality: Tools and Resources Turning a vision into reality isn't just about having a groundbreaking idea. It's about mobilizing the right tools and resources to bring that idea to life. This journey, from concept to execution, necessitates an array of elements—each crucial in transforming aspiration into tangible outcome. Let's delve into how one can navigate this path effectively, leveraging what's available and often, discovering new frontiers in the process.

The first step in this transformative journey is to solidify the concept itself. Ideation tools like mind mapping software or digital whiteboards can be instrumental in fleshing out ideas, allowing for a dynamic visualization of thoughts and potential pathways. Such tools encourage unrestricted brainstorming, crucial for the initial creative process.

Once the concept is clear and refined, the next phase is research and development (R&D). This is where digital libraries, online courses, and academic papers play an indispensable role. They provide access to existing knowledge and innovations, ensuring that one's own endeavor is not reinventing the wheel but rather, building upon or diverging from established paths with intention.

Prototyping comes next. Today, technologies such as 3D printing and computer-aided design (CAD) software have revolutionized prototyping, making it quicker and less costly. These resources allow for rapid iteration and testing of ideas, enabling creators to refine their concepts based on tangible feedback.

Yet, building something new isn't just about the product; it's also about building the right team. Collaboration tools, project management software, and digital communication platforms become essential in knitting together the collective genius of diverse teams, even if members are distributed across the globe.

Funding is another critical resource, often acting as the lifeline for turning visions into reality. Crowdfunding platforms, venture capital, angel investors, and grant opportunities represent various avenues through which projects can secure essential financial backing. Each of these requires a unique approach and understanding to effectively harness them.

Legal and regulatory guidance is equally important. As much as innovation is about breaking new ground, it must navigate the existing legal landscape. Legal advisory services and online resources focused on intellectual property, regulatory compliance, and business structuring are invaluable in this regard.

With a product and team in place, and legalities addressed, the focus shifts to market readiness. Digital marketing tools, social media platforms, and analytics software offer powerful means to build a brand, engage with audiences, and refine strategies based on real-time insights.

For operations, especially within the realm of the DIY factory model and New Work environments, lean management tools and methodologies help in streamlining processes, ensuring that creativity and innovation aren't bogged down by inefficiency.

Technology infrastructure plays a pivotal role as well. From basic needs like high-speed internet and reliable hardware to more specific requirements like cloud computing services, machine learning capabilities, and blockchain technology, the right tech stack can drive or hinder progress.

Education and continuous learning cannot be overlooked. Online platforms offering courses on everything from technical skills to business acumen and creative thinking empower individuals and teams to grow continually, staying ahead in a rapidly evolving landscape.

Community engagement is another potent tool. Engaging with local and global communities through events, workshops, and online forums fosters a culture of sharing and collaboration. It opens avenues for feedback, partnerships, and even friendships that fuel both personal and project growth.

Lastly, personal wellbeing tools, though often neglected, are crucial. Work-life balance apps, mindfulness and fitness apps, and time management tools ensure that the mind and body are in the best shape to innovate and create.

All these tools and resources, when leveraged efficiently, can collectively turn the daunting task of bringing an innovative vision to reality into an achievable, structured process. However, it's important to note that beyond these tangible resources, a few intangible ones—like resilience, adaptability, and vision—are equally vital. They propel one forward when challenges arise and ensure that the journey from vision to reality isn't just successful, but also enriching and transformative.

In conclusion, the pathway from vision to reality is multi-faceted and complex, yet incredibly rewarding. With the right mix of tools, resources, and intrinsic qualities, innovators can navigate this journey, transforming their bold visions into tangible realities that push boundaries, challenge the status quo, and ultimately, contribute to crafting a new era of work.

Building Partnerships and Networks

In the journey of transforming visionary concepts into tangible realities within the DIY factory and New Work ecosystem, forging strategic partnerships and networks is crucial. This chapter is dedicated to exploring how connections can be nurtured and leveraged to create a thriving environment for innovation. The essence of building a dynamic ecosystem lies in recognizing the potential of collective effort and diversity of thought.

Initiating this process requires a clear understanding of the ecosystem's objectives. Identifying key stakeholders - from suppliers and customers to collaborators and community leaders - is the first step towards building a network that supports and enriches the ecosystem. Effective networking goes beyond mere acquaintance; it demands recognition of mutual benefits and a shared vision for the future of work and manufacturing.

One of the foundational elements of establishing fruitful partnerships is trust. This involves developing relationships with stakeholders through transparency, reliability, and consistent communication. It's essential to convey your mission passionately and articulate how collaboration can drive mutual growth. Trust builds over time, laying a solid foundation for long-term collaborations.

Another critical aspect is identifying complementary strengths among potential partners. This can lead to synergies that enhance creativity, efficiency, and innovation. For instance, collaborations between tech startups and manufacturing experts can bridge the gap between cutting-edge technology and practical application. These partnerships can accelerate problem-solving and the development of new solutions.

Diversity in your network enriches the ecosystem with different perspectives, fostering innovative thinking. Engaging with a broad spectrum of industries, disciplines, and cultures can reveal unforeseen opportunities and solutions. This diversity also increases the resilience of the ecosystem, enabling it to thrive amidst challenges and adapt to change.

Active participation in industry-related forums, conferences, and online communities can serve as fertile ground for expanding your network. These platforms offer insights into emerging trends, technologies, and potential collaborators who share your enthusiasm for innovation and the DIY ethos. Moreover, contributing your expertise and resources can position you as a thought leader, attracting like-minded individuals and organizations to your network.

The advent of digital tools has revolutionized networking, allowing for the creation of virtual collaborations that transcend geographical limitations. Utilizing social media, professional networking sites, and collaborative platforms can significantly expand your reach, enabling you to connect with global innovators, thinkers, and creators. Digital networks also facilitate real-time communication and collaboration, essential for the fast-paced development cycles of DIY factories.

Developing a value proposition for your network is vital. This involves articulating how the partnership will deliver reciprocal value, whether through shared resources, joint ventures, or co-developed innovations. Clarity in the value exchange cements the partnership's foundation, ensuring that all parties are aligned with the collective goals.

As the network grows, fostering a culture of openness and collaboration becomes increasingly important. Encouraging partners to share knowledge, resources, and feedback can amplify the collective intelligence of the ecosystem. This culture of sharing and collaboration is the bedrock of a vibrant, innovative community that drives continuous improvement and breakthroughs.

Continuous engagement with your network is necessary to maintain its vitality and relevance. Regular updates, shared successes, and collaborative projects keep the community engaged and motivated. Moreover, celebrating milestones and achievements together strengthens bonds and renews the commitment to shared objectives.

It's also imperative to be open to evolving partnerships. As the ecosystem and its needs change, so too should the networks that support it. Being adaptable ensures that partnerships remain relevant and effective, fostering a dynamic environment that responds to new challenges and opportunities.

Mentorship within the network can play a transformative role, especially for emerging entrepreneurs and innovators. Seasoned professionals can offer guidance, share wisdom, and provide support to the next generation. This not only aids in their development but also ensures the longevity and evolution of the industry's ethos and values.

Finally, the success of these partnerships and networks should be measured and celebrated. Establishing metrics for collaboration, innovation, and impact can help quantify the benefits of the network. Celebrating these achievements not only acknowledges the collective effort but also inspires further innovation and collaboration.

In conclusion, building partnerships and networks is a strategic endeavor that demands dedication, foresight, and a commitment to shared goals. It's about creating a coalition of diverse talents and visions, united by the desire to innovate and redefine the landscape of work and manufacturing. Through trust, collaboration, and continuous engagement, the potential of the DIY factory and New Work ecosystem can be fully realized, shaping a future where innovation, creativity, and community thrive together.

The journey ahead may be complex and challenging, but the rewards of creating a connected, collaborative ecosystem are immense. This chapter has laid out the frameworks and strategies to cultivate meaningful partnerships and networks. Now, it's up to you to take these concepts and turn them into action. The future is collaborative, and by embracing these principles, you can build an ecosystem that not only survives but flourishes in the ever-evolving landscape of work and innovation.

Chapter 25:
The Role of Governments and Policy in Shaping the Future of Work

In navigating the transformative tides of the future of work, the role of governments and policy cannot be overstated. As we delve into the intricacies of how policy frameworks can either catalyze or stifle innovation and flexibility, it's essential to recognize that the creation of a more productive, creative, and fulfilling future lies at the intersection of visionary policy-making and grassroots innovation. The essence of this chapter is to explore the dynamic interplay between government initiatives and the burgeoning movements of DIY factories and New Work, underscoring the indispensable nature of policy in nurturing ecosystems conducive to the growth of modern work culture. Through a strategic blend of advocacy, public engagement, and the implementation of forward-thinking policies, governments have the potential to not only anticipate but also shape the evolving landscape of work. Such an approach calls for an unwavering commitment to fostering environments that empower both individuals and communities to thrive in the face of technological advancements and shifting economic paradigms. By examining case studies of national and local initiatives, we unveil the transformative power of policies designed with innovation, flexibility, and sustainability at their core, thereby laying the groundwork for a future where work is not just a means to an end but a canvas for human potential and societal progress.

Policy Frameworks Supporting Innovation and Flexibility

In a world where the pace of innovation often outstrips the ability of regulations to adapt, the onus falls on governments to craft policy frameworks that not only support but actively promote flexibility and innovation. Such frameworks are imperative for the nurturing of environments where DIY factories and New Work principles can not only emerge but thrive. By fostering an ecosystem that encourages experimentation, governments can unlock a wealth of creativity and economic potential, providing a solid foundation for future generations. This involves rethinking traditional regulatory approaches and embracing policies that facilitate easier access to funding, resources, and markets for startups and innovators. Furthermore, it requires the creation of educational and training programs that equip individuals with the skills needed to navigate and succeed in these new paradigms of work. By prioritizing policies that support flexibility, governments signal their commitment to fostering a dynamic, inclusive, and sustainable future of work. This shift not only enables the transformation of workplaces but also empowers individuals to lead more fulfilling and productive professional lives, ultimately contributing to a more vibrant and resilient economy.

Case Studies: National and Local Initiatives In the journey to reimagine our workspaces and cultures, it's crucial to examine real-world examples where new work principles and DIY factory ideologies have been implemented at both national and local levels. These initiatives not only provide a blueprint for success but also highlight the diversity of approaches that can be taken to reshape our professional environments.

At the national level, we see governments acknowledging the importance of modern work cultures and the DIY ethos in driving economic growth and innovation. For instance, countries like Finland and South Korea have introduced policies aimed at fostering entrepreneurship and creativity within the workforce. Finland's innovation system, Tekes, notably supports startups and projects that emphasize new work principles, such as flat hierarchies, open communication, and continuous learning. Similarly, South Korea's creative economy initiative aligns closely with the ideals of the DIY factory, aiming to harness the creativity and innovation of its populace to drive economic development.

Local initiatives often spring from community needs and the desire for a more personal and direct impact. Coworking spaces and makerspaces are prime examples of DIY factory and new work concepts taking root on a local scale. These spaces not only provide the tools and technologies necessary for innovation but also foster a sense of community among like-minded individuals. An exemplary case is the Brooklyn Navy Yard in New York City, which has been transformed into a hub for makers, artisans, and small businesses, providing an environment where creativity and collaboration flourish.

In Denver, Colorado, the RiNo (River North) Art District has become a beacon for artists and entrepreneurs who embody the DIY ethos. Through a combination of public and private initiatives, RiNo has cultivated an ecosystem where new work practices thrive, showcasing how local economies can be revitalized through creativity and innovation.

Education plays a pivotal role in embedding new work and DIY principles early on. The High Tech High network of schools in California integrates project-based learning with real-world applications, preparing students for a future where creativity, collaboration, and adaptability are key. This educational model acts as a microcosm of the DIY factory, emphasizing hands-on learning and problem-solving.

In Canada, the Toronto Tool Library is an excellent illustration of how DIY culture can be promoted at the community level. By providing access to tools and resources, the library lowers the barriers to creation and innovation, embodying the spirit of resource sharing and collaboration central to new work ideologies.

Meanwhile, on the corporate front, companies like Gore & Associates have long been champions of flat organizational structures and team autonomy, embodying the essence of new work principles. Gore's lattice organization structure eschews traditional hierarchies, encouraging innovation and direct communication among team members.

Another inspiring case is the city of Chattanooga, Tennessee, which leveraged its status as the first city in the Western Hemisphere to offer gigabit-speed internet citywide. This initiative not only catapulted Chattanooga into the future of technology and innovation but also transformed it into a thriving center for startups and new work methodologies.

In Scandinavia, Sweden's experimentation with six-hour workdays underscores a commitment to rethinking work-life balance, an integral aspect of new work principles. This bold initiative seeks to improve productivity and happiness by acknowledging the importance of personal time and wellness.

Back in Asia, Singapore's Smart Nation initiative embodies the fusion of technology with new work practices. By leveraging digital technologies, Singapore aims to enhance the living and working environments of its citizens, creating an ecosystem that supports innovation, collaboration, and continuous learning.

In summary, these case studies exemplify the multifaceted nature of implementing DIY factory and new work concepts. From government policies to local community projects, and educational reforms to corporate restructuring, the path toward innovative and fulfilling work cultures is as diverse as it is promising. Each case offers valuable insights into the practical application of these principles and serves as a source of inspiration for those seeking to pioneer change in their own spheres.

What stands out across these examples is the common thread of empowerment, community, and innovation. By embracing new work and DIY factory principles, entities at all levels can contribute to a future where work is not just a means to an end but a source of fulfillment and growth. The journey is ongoing, and these case studies represent the vanguard of change in our understanding and approach to work in the 21st century.

The implications of such initiatives are profound, raising essential questions about how we value work, creativity, and collaboration in the modern era. As we look to the examples set by these national and local initiatives, it becomes clear that the future of work is not a distant ideal but a present reality being shaped by those willing to innovate and advocate for change.

Yet, the road is not without challenges. These case studies also highlight the obstacles faced when introducing new paradigms—resistance to change, the need for new skills, and the importance of creating supportive ecosystems. But with persistence and a shared vision, the potential for transformative impact is immense.

In conclusion, as we forge ahead, let's draw inspiration from these national and local initiatives that have dared to reimagine the landscape of work. They remind us that with creativity, collaboration, and an unwavering commitment to innovation, we can build work cultures that not only drive economic growth but also nurture the human spirit. Let these examples light our path as we embark on this exciting journey toward a new era of work.

Advocacy and Public Engagement

In the realm of transforming work cultures and innovative practices, governments and policymakers play a critical role not only in the direct shaping of policies but also in catalyzing broader public engagement and advocacy efforts. At the heart of any significant societal shift lies the power of public awareness and the collective push for change. In the context of fostering the future of work, especially with the DIY factory model and New Work principles, advocacy, and public engagement serve as the bridge between vision and reality.

Advocacy efforts, when thoughtfully executed, can illuminate the myriad ways in which the DIY factory and New Work movement stand to benefit not just individuals but society at large. By illustrating the potential for greater creativity, innovation, and personal fulfillment, advocates can inspire a wider embrace of these concepts. Such advocacy is most effective when it transcends the limits of professional circles and engages a diverse range of stakeholders, including educators, students, entrepreneurs, and even those who may not currently see themselves as directly impacted by shifts in work culture.

Public engagement, on its part, involves forging deeper, more meaningful connections between the initiatives at the heart of the new work movement and the broader community. This engagement can take many forms, from workshops and public talks to interactive experiences that give individuals a hands-on understanding of the benefits of DIY factories and New Work principles. By making the abstract tangible, public engagement efforts demystify the innovative practices at play, fostering a sense of ownership and excitement among the broader population.

This dual approach of advocacy and public engagement necessitates a strategic use of various platforms and mediums. Social media, for instance, provides an unparalleled opportunity to reach vast audiences quickly and interactively. Traditional media, public forums, and educational institutions also serve as vital arenas for disseminating ideas and fostering dialogues that challenge the status quo.

Government and policy bodies have a unique leverage point in this process: the ability to not only endorse but actively promote these dialogues and engagements through policy frameworks and public programs. By providing platforms for discussions, funding public initiatives, or even facilitating partnerships between the private sector and community organizations, government bodies can amplify the impact of advocacy and public engagement manifold.

A notable example of successful public engagement can be seen in initiatives that showcase the tangible results of DIY projects and New Work environments. Open house events at innovation hubs or DIY factories, for instance, allow people to see firsthand the potential for creativity, collaboration, and productivity that these new models offer. Such initiatives not only educate but also excite the public, building grassroots support for wider adoption of these practices.

However, for advocacy and public engagement to truly resonate, they must also address the challenges and concerns inherent in transitioning to new work paradigms. Honest dialogues about the learning curves, the need for skill development, and the shifting nature of job security help in grounding the conversation in reality. It's essential for advocacy efforts to acknowledge these challenges, offering constructive paths forward rather than dismissing concerns.

Moreover, engaging with skeptics and critics forms a crucial part of advocacy and public engagement. By inviting open dialogue and debate, proponents of the DIY factory and New Work movements can further refine their visions and strategies. This transparent and inclusive approach not only strengthens the movement's foundations but also broadens its appeal.

Additionally, effective advocacy and public engagement hinge on showcasing diverse success stories that illustrate the broad applicability and benefits of DIY and New Work practices across different sectors and communities. Highlighting a range of case studies — from revitalized small businesses to empowered community groups — paints a vivid picture of potential and invites individuals and organizations to imagine how they might contribute to or benefit from this shift.

Ultimately, advocacy and public engagement are about building a movement — a collective push towards a future where work is more adaptable, creative, and fulfilling. This movement thrives on the active participation of individuals across the spectrum of society, each bringing their own insights, experiences, and aspirations to the table.

Such a broad-based, inclusive approach to advocacy and public engagement not only accelerates the adoption of new work models but also ensures that they are shaped by a diverse range of voices. This inclusivity is key to creating work environments that are not only innovative and productive but also equitable and responsive to the needs of all participants.

In driving this movement forward, patience and persistence are crucial. Significant societal shifts do not happen overnight, nor without setbacks. Yet, with a clear vision, strategic engagement, and a commitment to open dialogue, the journey towards reimagined work cultures becomes not just feasible but invigorating.

As we strive towards this future, let us remember that the role of advocacy and public engagement is not merely to persuade but to empower. By involving individuals and communities in the conversation, by providing them with the tools and knowledge to make informed decisions about the future of work, we are fostering a more dynamic, innovative, and resilient society. A society ready not just to adapt to the changes ahead but to lead them.

Therefore, let our efforts in advocacy and public engagement be characterized by creativity, inclusivity, and a steadfast belief in the transformative power of collective action. Together, we can shape a future of work that embodies our highest aspirations for innovation, equity, and fulfillment.

Chapter 26:
Envisioning a Productive, Creative, and Fulfilling Future

In the tapestry of the modern work culture, we are witnessing a remarkable transformation. This change is a testament to the power of innovation, the DIY movement, and the principles of New Work, knitting together a future that promises productivity, creativity, and fulfillment in abundance. The journey through the emerging landscapes of DIY factories and New Work paradigms has illuminated paths toward redefining what work means and how it can be a source of joy, growth, and community.

The emergence of DIY factories exemplifies a shift toward hands-on innovation and creativity, making the process of creating as rewarding as the outcomes. This approach embodies the essence of craftsmanship, where the value lies not only in the products but also in the mastery and personal growth achieved through the act of making. By embedding these principles in our professional environments, we foster a culture where innovation thrives, and every individual feels empowered to bring their unique ideas to life.

New Work principles further elevate this landscape by reshaping organizational cultures and structures. The move toward flat hierarchies and open communication channels underscores a commitment to inclusivity and collaboration. By valuing every team member's input and fostering an environment where autonomy and accountability go hand in hand, we unlock unprecedented levels of creativity and motivation. This transformation paves the way for workplaces that are not only more productive but also more humane and fulfilling.

The synergy of DIY factories and New Work concepts does not merely forecast a future of work that is different from what we know today; it actively constructs it. Through case studies and best practices shared throughout this exploration, we've seen the tangible impacts of these paradigms. Companies and communities that have embraced these principles report not just improved outcomes but a revitalized sense of purpose and engagement among their members.

This shift also heralds a redefinition of leadership roles. In this new era, leaders are facilitators and visionaries who inspire innovation and creativity. They lead by example, empower their teams, and navigate the complexities of change with empathy and resilience. Their role is pivotal in crafting environments where imagination flourishes, and challenges are viewed as avenues for learning and growth.

Embedded within these discussions is an undeniable emphasis on technology as a crucial enabler. From automating mundane tasks to leveraging AI for creative problem-solving, technology amplifies our capacity to innovate. However, it's the human touch—the artistry and ingenuity of individuals—that steers these tools toward meaningful and transformative outcomes.

In contemplating the economic impact, it becomes clear that DIY factories and New Work principles are not just about individual or organizational gains. They represent a shift toward a more production-oriented society, where value creation overtakes mere consumption. This model not only promises economic resilience but also advocates for a more equitable distribution of wealth and opportunities.

Yet, as we stand on the brink of this promising future, we also recognize the barriers and challenges that lie in its realization. Skepticism, resistance to change, and the inertia of established norms pose significant obstacles. Overcoming these requires perseverance, a commitment to continuous learning, and an open-minded approach to experimentation and adaptation.

Driving this transformation requires not just visionary leaders but also a community of practitioners and advocates. A global network of individuals and organizations committed to these principles can amplify their impact, catalyzing broader societal shifts. This collective effort can redefine not just how we work but also how we live, learn, and connect with each other.

The role of education and skill development cannot be overstated in paving the way for this future. Rethinking education to nurture creativity, critical thinking, and adaptability is fundamental. By aligning educational systems with the needs of this evolving work landscape, we prepare future generations to thrive and shape their destinies.

In addition, policy frameworks and government support play a crucial role in enabling these changes. Policies that encourage innovation, flexibility, and collaboration can provide the necessary groundwork for DIY factories and New Work principles to flourish. By addressing legal, ethical, and regulatory considerations proactively, we can create an ecosystem that supports sustainable and inclusive growth.

The financial models and resources driving sustainability and growth within this realm also demand attention. Crowdfunding, venture capital, and other funding avenues offer the lifeline for these initiatives to take off and scale. Equally important is cultivating economic resilience and diversification to withstand challenges and seize emerging opportunities.

As we chart this path forward, the role of personal transformation and entrepreneurial mindset emerges as a cornerstone. Embracing change, learning from failures, and balancing passion with pragmatism are vital for anyone aspiring to lead or contribute to this movement. It's about cultivating an inner resilience and openness to new experiences that fuel innovation and growth.

Finally, the integration of art, culture, and work underscores the holistic nature of this transformation. Encouraging creative expression and cultural engagement within the work environment enriches our lives, fosters community, and enhances our capacity for empathy and innovation. It's a reminder that at the heart of all technological advancement and organizational change lies human creativity and connection.

In envisioning a productive, creative, and fulfilling future, we are not just imagining possibilities; we are laying the foundation for a reality where work enriches our lives and societies. As we step into this future, let us carry forward the lessons, principles, and inspirations from the realms of DIY factories and New Work. Together, we can shape a world where innovation, creativity, and fulfillment are not just ideals but the cornerstones of our daily lives and work.

Appendix A

In the pursuit of pioneering a future where DIY factories and New Work principles shape our professional environments, knowledge equips us not just to participate in the conversation but to lead it. This appendix gathers a curated list of resources for those ready to delve deeper into the transformative practices that lie at the intersection of innovation, culture, and the DIY movement.

Resources for Further Exploration

Books and Publications: Expand your understanding and draw inspiration from a spectrum of authors who have charted the course of DIY and New Work movements. Their insights provide a foundation upon which to build your unique perspective and approach.

Online Platforms and Forums: Join the thriving online communities where ideas flourish through sharing and collaboration. Engage in discussions, pose questions, and connect with like-minded individuals who are equally passionate about shaping the future of work.

Workshops and Conferences: Immersive experiences facilitate profound learning. Seek out events that offer hands-on sessions or thought-provoking talks by pioneers in the field. These gatherings provide networking opportunities and the chance to witness innovation in action.

Toolkits and Guides: Practical resources can guide you in transforming theory into action. From setting up your DIY factory to implementing New Work structures, these toolkits offer step-by-step instructions and best practices.

Case Studies: Learn from the successes and challenges faced by those who have embarked on similar journeys. Analyzing case studies provides valuable insights and can inspire strategies that could be adapted to your context.

Innovation isn't just about what you do; it's about how you think. Each resource listed above is a beacon, guiding towards not just the adoption of new practices but towards a profound shift in mindset. The journey ahead is as much about personal growth as it is about professional development.

Embrace this adventure with openness, curiosity, and the willingness to question the status quo. It's through exploration and continuous learning that we can truly become architects of the future.

Resources for Further Exploration

In our journey through the transformative landscapes of DIY factories and New Work environments, we've uncovered the layers that compose these innovative practices. Yet, the exploration doesn't end here. For those inspired to dig deeper and perhaps embark on their own ventures into this shifting paradigm, a wealth of resources awaits. This concluding section of our appendix is dedicated to guiding you toward a multitude of platforms, publications, and communities that can serve as your compass in this adventure of continuous learning and experimenting.

Books are foundational in understanding the complex ideologies and methodologies that underpin the DIY factory and New Work concepts. I recommend starting with titles that delve into the history and philosophy of the maker movement, as well as those that explore the future of work. These readings not only provide a solid theoretical background but also showcase practical case studies of successful implementations of these models across various industries.

Online courses and workshops present an accessible way to acquire specific skills or deepen your understanding of particular aspects of DIY factories and New Work practices. Platforms like Coursera, Udemy, and FutureLearn offer courses ranging from project management in innovative environments to the technical skills needed to operate digital fabrication tools. Participating in such courses allows for an experiential approach to learning, often giving you the ability to implement small-scale projects as you progress.

Following influential bloggers and thought leaders in the space can offer both inspiration and practical advice. These individuals often share their journeys, the challenges they've faced, and how they've overcome them, providing a personal touch to the learning experience. Their insights can be invaluable in avoiding common pitfalls and sparking new ideas for your own projects.

Membership in online forums and social media groups dedicated to DIY manufacturing and New Work can also be incredibly beneficial. These communities are spaces for sharing knowledge, asking questions, and connecting with like-minded individuals. Platforms like Reddit, LinkedIn groups, and specialized online forums host vibrant discussions on everything from the latest in 3D printing technology to strategies for fostering innovation within a traditional corporate structure.

Conferences and meetups are exceptional places to immerse yourself in the current state of the art, network with pioneers in the field, and even find potential collaborators. Events such as Maker Faire, the Work Awesome conference, and local DIY factory open houses provide opportunities to see firsthand the innovations shaping the future of work and manufacturing.

For those interested in the technological side, following the development of open-source projects and platforms can be enlightening. These resources often include comprehensive documentation and community support, making them an ideal starting point for implementing technology-driven projects within your own DIY factory or New Work initiative.

Podcasts offer another avenue for exploration, with series dedicated to every aspect of innovation, from design thinking to the nuts and bolts of running a maker space. They provide the convenience of learning on the go and can be a great source of motivation, offering regular doses of stories from the forefront of the New Work movement.

Documentaries and video series on platforms like YouTube or Vimeo cater to visual learners and those curious about the day-to-day realities of innovative workspaces around the globe. These resources often provide a behind-the-scenes look at successful DIY factories and entrepreneurial ventures, highlighting both the challenges faced and the solutions found.

Professional associations related to your specific interest within the DIY and New Work realms can provide a structured pathway for exploration and growth. Whether it's an association focused on robotics, sustainable manufacturing, or innovative business models, membership can offer access to exclusive resources, industry insights, and professional development opportunities.

Finally, exploring local initiatives and workspaces can provide a tangible sense of what's possible. Many cities now boast collaborative workspaces, innovation hubs, and maker labs open to the public. Engaging with these local resources not only supports the community but also connects you to a network of potential mentors, partners, and friends who share your passions and interests.

The ecosystem surrounding DIY factories and New Work is constantly evolving, with new tools, techniques, and philosophies emerging at a rapid pace. Keeping abreast of the latest developments requires a commitment to ongoing education and an openness to experimentation. Whether you're just beginning your journey or looking to deepen your existing practice, the resources mentioned here can catalyze your efforts, providing both inspiration and practical guidance.

As we close this chapter, remember that the transformation towards more innovative, inclusive, and personally fulfilling ways of working is a collective endeavor. Each book read, course taken, and community joined not only propels you forward but contributes to the broader movement reshaping our understanding of what it means to work and create. Let this exploration be not just an end but a beginning, a launchpad into a future we build together.

Embark on this journey with an open mind and a readiness to experiment, learn, and ultimately contribute your unique voice to the chorus of innovation that is defining the new era of work and manufacturing. The path may be challenging, but the rewards—personal growth, community impact, and the joy of creation—are immeasurable. Let's explore the vast landscapes of possibility that lie ahead, equipped with the knowledge that we are all architects of the future we wish to see.